GUIMAROTA
A Jurassic Ecosystem

Impressum

Die Deutsche Bibliothek - CIP-Einheitsaufnahme

Guimarota : a jurassic ecosystem /
Thomas Martin & Bernard Krebs (ed.). -
München : Pfeil, 2000
  ISBN 3-931516-80-6

Printed with financial support by the
Deutsche Forschungsgemeinschaft (DFG)

Front cover:
Henkelotherium guimarotae KREBS 1991

Back cover:
Tunnel in the Guimarota mine

Frontispiece:
The Late Jurassic Guimarota Ecosystem. Drawing by *M. BULANG-LÖRCHER* after a draft of *T. MARTIN*

Copyright © 2000
Verlag Dr. Friedrich Pfeil, München, Germany

Druckvorstufe: Verlag Dr. Friedrich Pfeil, München
CTP-Druck: grafik + druck GmbH Peter Pöllinger, München
Buchbinder: Thomas, Augsburg

ISBN 3-931516-80-6

Printed in Germany

Printed on permanent and durable acid-free paper

Verlag Dr. Friedrich Pfeil
Wolfratshauser Straße 27
D-81379 München, Germany
Tel. ++49-89-74 28 27 0 • Fax ++49-89-72 42 772
e-mail: 100417.1722@compuserve.com

GUIMAROTA

A Jurassic Ecosystem

Thomas MARTIN & Bernard KREBS
(Editors)

Verlag Dr. Friedrich Pfeil · München 2000
ISBN 3-931516-80-6

Authors

PD Dr. Martin Aberhan, Institut für Paläontologie, Museum für Naturkunde der Humboldt-Universität zu Berlin, Invalidenstraße 43, D-10115 Berlin;
E-mail: martin.aberhan@rz.hu-berlin.de

Dr. Annette Broschinski, Niedersächsisches Landesmuseum, Naturkunde-Abteilung, Willy-Brandt-Allee 5, D-30169 Hannover;
E-mail: ABroschin@compuserve.com

Ellen Drescher, Institut für Paläontologie, Freie Universität Berlin, Malteserstraße 74-100, D-12249 Berlin;
E-mail: palaeont@zedat.fu-berlin.de

Dipl.-Geol. Thomas Gassner, Institut für Paläontologie, Museum für Naturkunde der Humboldt-Universität zu Berlin, Invalidenstraße 43, D-10115 Berlin;
E-mail: tgassner@debitel.net

Dipl.-Geol. Uwe Gloy, Institut für Paläontologie, Freie Universität Berlin, Malteserstraße 74-100, D-12249 Berlin;
E-mail: dasgloy@zedat.fu-berlin.de

Prof. em. Dr. Gerhard Hahn, Philipps-Universität, Fachbereich Geowissenschaften, Hans-Meerwein-Straße, D-35032 Marburg

Dr. Renate Hahn, Berliner Straße 31, D-35282 Rauschenberg;
E-mail: Rhahn77234@aol.com

PD Dr. Rolf Kohring, Institut für Paläontologie, Freie Universität Berlin, Malteserstraße 74-100, D-12249 Berlin;
E-mail: palaeont@zedat.fu-berlin.de

Prof. a.D. Dr. Bernard Krebs, formerly Institut für Paläontologie der Freien Universität Berlin; current address: 60, Grande Rue, Chanceaux, F-21440 St. Seine-l´Abbaye, France

Elisabeth Krebs, 60, Grande Rue, Chanceaux, F-21440 St. Seine-l'Abbaye, France

Dr. Jürgen Kriwet, Institut für Paläontologie, Museum für Naturkunde der Humboldt-Universität zu Berlin, Invalidenstraße 43, D-10115 Berlin;
E-mail: Jkriwet@hotmail.com

PD Dr. Thomas Martin, Institut für Paläontologie, Freie Universität Berlin, Malteserstraße 74-100, D-12249 Berlin;
E-mail: tmartin@zedat.fu-berlin.de

Dr. Barbara Mohr, Institut für Paläontologie, Museum für Naturkunde der Humboldt-Universität zu Berlin, Invalidenstraße 43, D-10115 Berlin;
E-mail: barbara.mohr@rz.hu-berlin.de

Manuela Nowotny, Institut für spezielle Zoologie und Evolutionsbiologie, Friedrich-Schiller Universität, Erbertstraße 1, D-07743 Jena;
E-mail: manu_mucha@yahoo.de

Dr. Oliver Rauhut, Institut für Paläontologie, Freie Universität Berlin, Malteserstraße 74-100, D-12249 Berlin;
E-mail: oliver.rauhut@rz.hu-berlin.de

PD Dr. Frank Riedel, Institut für Paläontologie, Freie Universität Berlin, Malteserstraße 74-100, D-12249 Berlin;
E-mail: palaeont@zedat.fu-berlin.de

PD Dr. Michael Schudack, Institut für Paläontologie, Freie Universität Berlin, Malteserstraße 74-100, D-12249 Berlin;
E-mail: schudack@zedat.fu-berlin.de

Dr. Stephan Schultka, Institut für Paläontologie, Museum für Naturkunde der Humboldt-Universität zu Berlin, Invalidenstraße 43, D-10115 Berlin;
E-mail: stephan.schultka@rz.hu-berlin.de

Dipl.-Geol. Daniela Schwarz, Institut für Paläontologie, Freie Universität Berlin, Malteserstraße 74-100, D-12249 Berlin;
E-mail: crocschwarz@hotmail.com

Dipl.-Geol. Marc Filip Wiechmann, Institut für Paläontologie, Freie Universität Berlin, Malteserstraße 74-100, D-12249 Berlin;
E-mail: marcfilip@hotmail.com

Contents

Preface		7
1	The excavations in the Guimarota mine BERNARD KREBS	9
2	Geological setting and dating of the Guimarota beds MICHAEL E. SCHUDACK	21
3	The flora of the Guimarota mine BARBARA MOHR & STEPHAN SCHULTKA	27
4	Ostracodes and charophytes from the Guimarota beds MICHAEL E. SCHUDACK	33
5	The mollusk fauna from the Guimarota mine MARTIN ABERHAN, FRANK RIEDEL & UWE GLOY	37
6	The fish fauna from the Guimarota mine JÜRGEN KRIWET	41
7	The albanerpetontids from the Guimarota mine MARC FILIP WIECHMANN	51
8	The turtles from the Guimarota mine THOMAS GASSNER	55
9	The lizards from the Guimarota mine ANNETTE BROSCHINSKI	59
10	The crocodiles from the Guimarota mine BERNARD KREBS & DANIELA SCHWARZ	69
11	The dinosaur fauna from the Guimarota mine OLIVER W. M. RAUHUT	75
12	Pterosaurs and urvogels from the Guimarota mine MARC FILIP WIECHMANN & UWE GLOY	83
13	Eggshells from the Guimarota mine ROLF KOHRING	87
14	The docodont *Haldanodon* from the Guimarota mine THOMAS MARTIN & MANUELA NOWOTNY	91
15	The multituberculates from the Guimarota mine GERHARD HAHN & RENATE HAHN	97
16	The dryolestids and the primitive "peramurid" from the Guimarota mine THOMAS MARTIN	109
17	The henkelotheriids from the Guimarota mine BERNARD KREBS	121
18	Taphonomy of the fossil lagerstatte Guimarota UWE GLOY	129
19	Preparation of vertebrate fossils from the Guimarota mine ELLEN DRESCHER	137
20	Overview over the Guimarota ecosystem THOMAS MARTIN	143
21	Bibliography of the Guimarota mine ELISABETH KREBS	147
Appendix: Floral and faunal list of the Guimarota mine		153

Cross references in this volume are cited as "2000*".

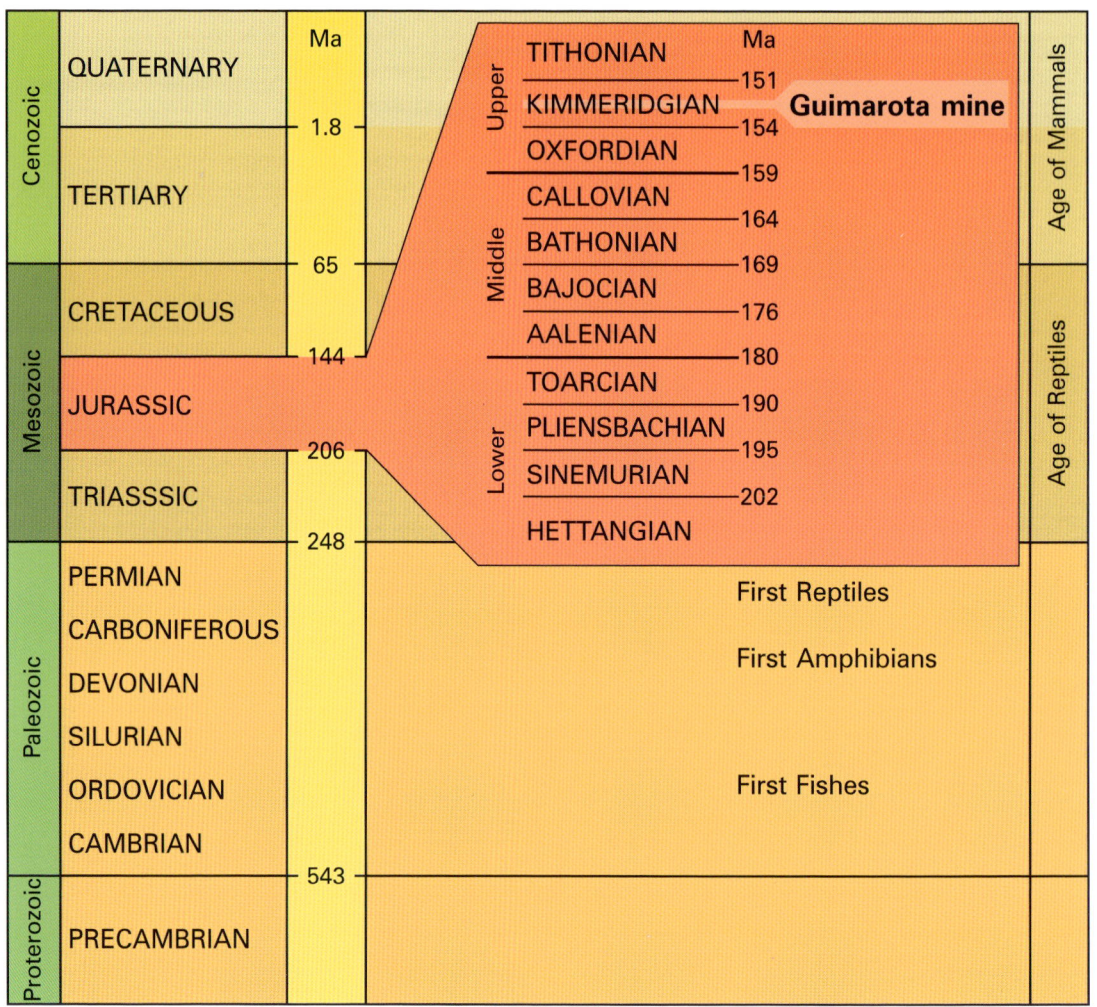

The setting of the Guimarota mine in earth history within the Mesozoic, the "age of reptiles". Drawing by UWE GLOY.

Preface

The evolution of life on earth, which began more than three billion years ago, is a unique historical process. It cannot be recapitulated in laboratory experiments, or recalculated on a computer. Only fossils, the remains of plants and animals of former times, can provide us with information about it. The science of paleontology, the historical branch of the biological sciences, investigates the evolution of the biosphere of the earth. During the last two hundred years, paleontologists have developed a more or less secured picture of this constantly changing living world. However, many gaps remain in our understanding of these events. Especially the origin and early stages of the evolution of several large and important groups of animals and plants are still in the dark. The beginnings of mammals is an example that affects us in particular, since it concerns our own ancient roots, but the early history of other vertebrate groups, like that of modern amphibians (frogs and salamanders) and lizards and snakes, is also still poorly understood. We know their distant ancestors quite well, which were diverse and widely distributed in the Late Paleozoic ("ancient life"; the era of fishes and amphibians) and the Triassic, the oldest period of the Mesozoic ("middle life"; the era of reptiles). These ancestors are the stegocephalians (primitive amphibians with a complete skull roof) in the case of amphibians, basal diapsids (reptiles with two pairs of openings in the posterior part of the skull roof) in the case of lizards and snakes, and the so-called mammal-like reptiles in the case of the mammals. These generally rather large ancestral taxa disappear at the end of the Triassic with few exceptions, some of which are still alive today as "living fossils", such as the tuatara in New Zealand. After the Triassic, there is a long period with little fossil evidence of the evolution of these lineages in Earth history. This gap in the fossil record spans the Jurassic and most parts of the Cretaceous, thus a period of some one hundred and fifty million years. Only at the end of the Cretaceous, the fossil record becomes better again, apart from the recently discovered extraordinary fossil localities in the province of Liaoning in north-eastern China. The mammals in particular experience a radiation from the latest Cretaceous on onwards, which is rapid according to geological time scales, so that it is often referred to as an "explosive" radiation, although it took several million years.

The gap mentioned above is based on a poor fossil record, since there must have been a continuous evolution from the ancestors to the successful descendants of the Cenozoic. Moreover, separate and often fragmentary remains of these transitional forms have been found. The poor fossil record of frogs, salamanders, lizards and snakes, and mammals in the Jurassic and Cretaceous can be explained as follows:

The animals are very small and furthermore lived on dry land, where their chances of becoming fossilized are much lower than those of aquatic organisms. The latter sink to the bottom of the oceans, lakes or rivers, where they are rapidly buried by sediment and are thus preserved. On the dry land, the corpses are usually unprotected by sedimentary covering and thus subject to destruction by scavengers and weathering. Only if they are washed into a body of water – rivers, lakes, lagoons, near-shore marine waters – they stand a chance of preservation. In this respect, swamps are also important, where incompletely decayed plant remains slowly turn into brown coal in stagnant, oxygen-poor waters. Corpses that happen to get into these hostile environments are usually safe from scavengers and no parts of the body are transported away. Another special case in the preservation of terrestrial animals are fissure fillings that form on calcareous plateaus. Often thousands of jaw- and skeleton remains are found in the clay that fills the fissures. However, mainly isolated bones and teeth are found in these localities, since the skeletons are usually completely disarticulated.

Europe – the best known area of the world in terms of paleontology – was largely covered by oceans in the Jurassic and Cretaceous. The famous European vertebrate localities from this era, such as Holzmaden and Solnhofen, contain an almost entirely marine fauna. Only during geologically short intervals, small areas of Europe emerged and formed islands of dry land.

As mentioned above, the ancestors of mammals and other modern terrestrial vertebrates were small animals. Large animals are specialized in terms of their large size alone; they have a long evolutionary history and generally are final forms. Remains of such large animals (e.g. dinosaur bones) are found by chance, they attract the attention of quarry workers, geologists, or interested laymen. The tiny bones and teeth of amphibians, lizards and small mammals, however, cannot be discovered without special equipment.

The paleontologist WALTER GEORG KÜHNE, the founder of the professorship and later the Institut für Paläontologie at the Freie Universität Berlin, was the first to dedicate his work to the systematic search for evidence concerning the early evolution of mammals. This was only part of his inclination to use theoretical considerations to discover objects and materials that are difficult to find, including amber and gold. Until then, paleontologists had only worked with chance discoveries, or carried out excavations or expeditions in areas that had yielded remarkable fossils by chance before, like Wyoming (USA), or the Gobi desert (Mongolia).

Being aware of the reasons for the poor documentation of Mesozoic mammals given above, W. G. KÜHNE began prospecting. First, Jurassic or Cretaceous fresh-water deposits had to de found. Even though such deposits do not have wide distributions in Europe – work in other parts of the world was out of the question in the 1950ies – the search had to be restricted to the most promising areas. KÜHNE thus used so-called indicator faunas, especially finds of dinosaurs, which were published or listed in museum catalogues. Early mammals, contemporaneous with dinosaurs, and, like the latter, inhabitants of terrestrial environments, were consequently found in several of these localities. In accordance with the small size of the objects he was looking for, methods that were devised for micropaleontological research had to be used.

Fissure fillings also attracted the attention of W. G. KÜHNE. First successes in southern England and Wales, though, could not be repeated in continental Europe. However, KÜHNE paid special attention to brown coals, which he knew as vertebrate localities from the – admittedly Tertiary – coals of the Geiseltal from his times as a student in Halle/Saale. He knew from his experience that there was the possibility to find not only isolated bones and teeth, but more complete remains, like skulls or even complete skeletons in brown coals.

However, in this type of localities, no indicator faunas were to be expected; never had dinosaur bones been found in Mesozoic brown coals. Therefore, KÜHNE had to visit all Jurassic and Cretaceous brown coal localities in Europe. He found the information he needed to do this in the geological literature, or in talks with field geologists. It was in the framework of these prospecting activities that he discovered the coal mine of Guimarota.

Acknowledgments

The excavations in Guimarota were generously funded by the Deutsche Forschungsgemeinschaft (German Research Foundation) and the Freie Universität Berlin; we are deeply indebted to both institutions. Thanks are also due to the Serviços Geológicos de Portugal and its directors – ANTÓNIO DE CASTELO BRANCO, FERNANDO MOITINHO DE ALMEIDA und MIGUEL M. RAMALHO – for giving the permissions for the fieldwork and their goodwill towards our work.

Cordial thanks are due to our Portuguese coworkers, especially to the foreman GUILHERME FERNANDOS DOS SANTOS, his wife ENCARNAÇÃO, as well as their son GUILHERME and daughter CECÍLIA, but also to the miners – the excavation would have been impossible without their experience – and the female helpers, the names of whom are noted down in the lists of the finds.

The fossils from Guimarota were skillfully prepared by Mrs. ELLEN DRESCHER, the majority of the difficult photographic work was carried out by Mrs. PETRA GROSSKOPF, the artful drawings were executed by Mr. PETER BERNDT and Mrs. MONIKA BULANG-LÖRCHER; Mr. WOLFGANG MÜLLER helped with the work at the Scanning Electron Microscope and printed the photographs; Mrs. WALTRAUT HARRE, Museum für Naturkunde der Humboldt Universität zu Berlin, did the reproductions of large-scale drawings. The German text was translated by Oliver RAUHUT. All of the above earned our sincere thanks.

Berlin, December 15th 1999
The editors

The excavations in the Guimarota mine

Bernard Krebs

First excavations

The Iberian Peninsula was especially suited for W. G. KÜHNEs plan to methodologically search for Mesozoic mammals, since large areas of the peninsula formed dry land from the Late Jurassic to the Early Cretaceous. First explorations started in the beginning of the 1950ies at the southern rim of the Pyrenees. In 1957, A. F. DE LAPPARENT & G. ZBYSZEWSKI published a monograph on the dinosaurs from Portugal, which drew KÜHNEs attention to this country. In September and October 1959, he traveled to the Portuguese dinosaur localities, together with a student of geology, W. FREY. In the course of this journey, he learned that there was a coal mine close to the town of Leiria, which was still working; the last in the Lusitanian basin. This was the mine of Guimarota, where Jurassic brown coals were still being mined (Figs. 1.1 and 1.2).

On his first visit, W. G. KÜHNE recognized shimmering white shells of clams, snails, and ostracodes on the surface of the recently mined coal, an indication that the coal formed in water that was rich in lime, which prevented the solution of vertebrate bones and teeth by humin acids. At the first examination of the coals, fragments of turtle shells and osteoderms of crocodiles were found, as well as scales of ganoid fishes. On October the 5th, 1959, W. FREY picked up a piece of coal that contained a fossil, and asked: "Professor, is this what you are looking for?" It was a skull remain of a multituberculate, an early mammal.

A first exploration campaign in the spring of 1960, in which W. G. KÜHNE was assisted by his wife URSULA and the students F. F. HELMDACH and G. KRUSAT, led to the discovery of further remains of early mammals. In September and October of the same year, the method for the investigation of the coal, which proved to be very successful in later years, was used for the first time, greatly assisted by the helpfulness of the capataz (pit foreman) of the mine, LUIZ FERNANDES: fist-sized and larger pieces were sorted out from the mined coal and split and examined by three local women. Apart from W. G. KÜHNE and his wife, the students G. KRUSAT, A. LIEBAU and F. SELIGMANN helped with this work. The number of mammalian remains discovered satisfyingly increased.

This success justified a longer field work season in 1961, from March to September. In the middle of April, KÜHNEs assistant S. HENKEL took over the technical lead of the excavation. He rationalized the proceedings of the work and employed more local helpers, which led to a further increase in the number of discoveries of vertebrate remains. However, a special difficulty arose: the mining company went bankrupt, most probably because the main customer for the coal, the cement factory in Leiria, changed to burning

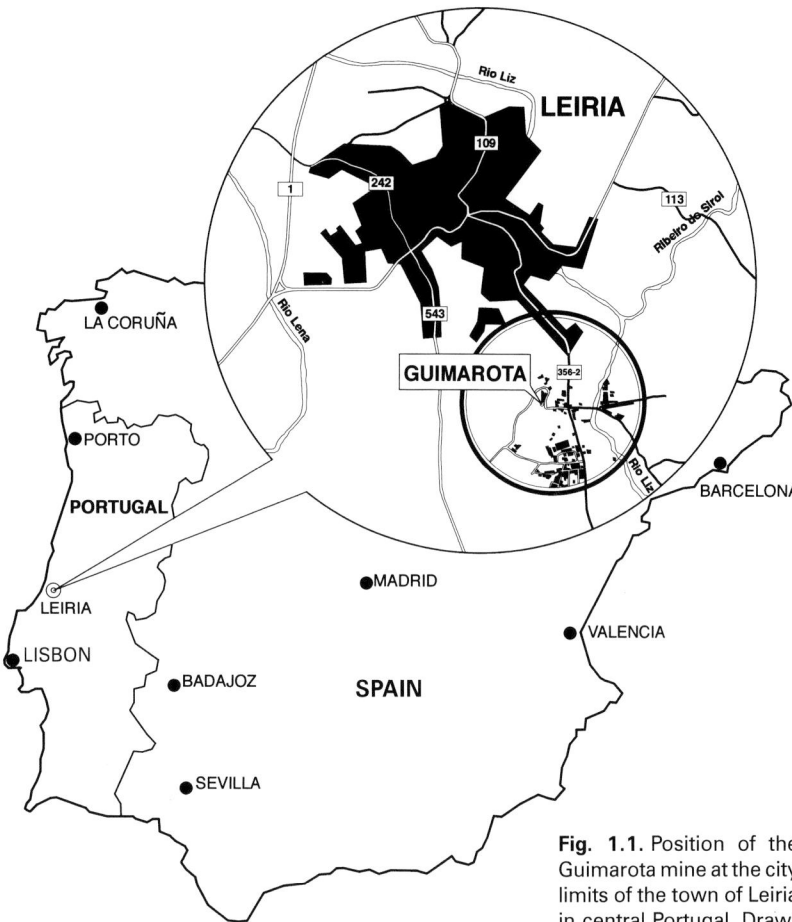

Fig. 1.1. Position of the Guimarota mine at the city limits of the town of Leiria in central Portugal. Drawing by UWE GLOY.

Fig. 1.2. Overview of the area of the mine, as it looked like during the old excavations at the beginning of the 1960ies (the photo was taken in 1967). On the western slope of the valley of the river Liz, above the roofs of the houses of the city district of Guimarota, the chimneys of two lime kilns, the long hall over the entrance of the mine, the tower for the transformers and the high building for the engines are distinguishable from left to right.

oil instead of coal. The mining stopped. A small amount of coal could be recovered with the help of some of the money that had been commissioned for the field work. Soon, however, the supply of fresh coal ceased completely. In the lack of new material, and in the assumption that only a part of the bone remains that were contained within the sediment were found by splitting the coal, S. HENKEL further processed the already examined coal by dissolving and screenwashing it. The thus concentrated sediment was further concentrated in Berlin and finally picked for fossils. Some additional isolated mammalian teeth were really found.

After that, work in the locality Guimarota continued from August to October 1962, this time just to process the coaly sediment from the slagheaps (Fig. 1.3). For this work, S. HENKEL developed a half-continuous water-processing and screening method, which proved to be very efficient and was later, often modified and optimized, also used in other localities. S. HENKEL published his method in 1966. In 1968, W. G. KÜHNE called this method the "HENKEL process", under which name HENKELs method became known since (see Fig. 1.12). This was the last field work carried out at the locality for the time being. Since 1961, the Berliners D. ANDRES, H. GOCHT, G. HAHN and his wife, H. NORTHE (the later Mrs. HENKEL), as well as F. NUTSCH and his wife also took part in the work. In the course of these first excavations in the Guimarota mine, F. F. HELMDACH worked on the geology of the mine and its surroundings (1966). From 1963 to 1968, only a few bags of material from the slagheaps were taken to other excavation sites on the Iberian Peninsula, to be processed there with the help of the equipment used at these localities.

The money for the field work described above was provided by the Deutsche Forschungsgemeinschaft (German Research Foundation) under the project "Mesozoic Mammals"; the total costs added up to approximately 60,000 Deutsche Mark, only 1,500 DM of which were spent for the first prospecting in 1959. W. G. KÜHNEs activities in the Guimarota mine were benevolently supported by the Geological Survey of Portugal, in particular by its director at that time, Dom ANTÓNIO DE CASTELO BRANCO.

The reasons that caused W. G. KÜHNE to stop work at such a successful excavation were the uncertain property situation of the mine campound after the bankruptcy of the company, and the assumption that the fauna from the coals of Guimarota had been sampled completely. In personal talks, however, he admitted that his aversion to routine work also played a role in this decision; he strongly disliked excavations that carried on over decades. KÜHNE and his cowork-

Fig. 1.3. Processing of material from the slagheaps in front of the old lime kilns, using the HENKEL-process.

Fig. 1.4. Overview over the area of the mine during the excavation campaign from 1972 to 1982 (the photograph was taken in March 1975). The lime kilns are gone, only two arches of the hall remain, the tower for the transformers and the engine building are unchanged. The new roof over the entrance of the mine is discernible on the left of the hall. The buildings are still surrounded by fruit plantations.

ers prospected new areas in the following time, hoping to find easier accessible and possibly even richer localities for Mesozoic mammals. However, these hopes were not fulfilled.

The early excavations in the Guimarota mine resulted in the discovery of 74 upper and lower jaws of mammals with dentitions of varied preservation – from fragments to almost complete skulls – and some 1,300 isolated mammalian teeth, as well as numerous remains of fishes, amphibians and reptiles. Thus, the Guimarota mine at that time already was the richest Jurassic mammal locality world-wide. In contrast to the great attention in the media that accompanies every new fossil find nowadays, and the considerable press interest that the approximately contemporaneous discovery of the early hominid *Oreopithecus* received, the news about the Guimarota mine hardly reached the experts, not to mention the public. W. G. KÜHNE only published three short communications in the News Bulletin of the Society of Vertebrate Paleontology and preliminary reports in German, English, and French, respectively, in scientific journals (KÜHNE 1961a,b,c). He reported on his excavations in the Guimarota mine at the meetings of the Paläontologische Gesellschaft (Paleontological Association) in 1961 and 1963. A detailed account of his activities in this locality was only published in 1968.

The new excavation

When I arrived in Berlin in my new position as an assistant to the newly created professorship of paleontology at the Freie Universität Berlin in January 1964, I was surprised and excited by the richness and quality of the fossil material from Guimarota, since, among others, it contained skulls of reptiles and mammals that were hitherto only known from dentitions. The material obviously represented a locality that was highly unusual for the Jurassic, the further exploration of which justified any effort in my eyes. However, W. G. KÜHNE was unwilling to change his decision to explore other localities.

After field work in the basin of Tremp (province of Lérida, Spain), which unfortunately proved to be unsuccessful, I drove to Portugal in September of the year 1964, to take a look at the Guimarota mine. All the buildings and equipment were still present in the estate of the disused mine, including the impressive steam engine which was used for the extraction of the coal. In Lisbon, I asked for an appointment with the new director of the Geological Survey of Portugal, Senhor FERNANDO MOITINHO DE ALMEIDA. He turned out to be a very friendly, open-minded and interested conversationalist, who assured me of his support for a possible resumption of the paleontological activities in the Guimarota mine. In 1968, after an excavation in the locality of Porto Pinheiro (close to Lourinhã, Portugal), I had the opportunity to discuss the conditions of a resumption of the work in the Guimarota mine with Senhor MOITINHO in detail.

Our activities at Porto Pinheiro were of limited success, as was the case with all the other prospecting work and excavations in the late 1960ies and 1970ies, and the material recovered never reached the level of informativeness of the material from Guimarota. If we found anything, it were either only isolated teeth of mammals – which might document the high diversity of Mesozoic mammals, but do not allow detailed investigations of the evolutionary history of this group – or poor mammalian faunas which consisted only of few taxa. Prospecting trips to Morocco and Persia turned out to be unsuccessful.

These experiences reinforced the opinion held by my colleagues and me that the coal of the Guimarota mine should be further explored, and we accordingly worked out a project. W. G. KÜHNE refused to give his approval for our endeavor: it would not yield any new information. Apart from that he had shifted his attention to completely different topics in the meantime. However, the new conditions resulting from the reform of the universities in 1968 allowed us to proceed with the Guimarota project without the professor.

To clarify the conditions at the locality, S. HENKEL traveled to the Guimarota mine in the summer of 1971, together with G. KRUSAT and the student U. MÜRKENS. On the area of the former mine, they found a newly planted and blooming peach tree plantation. All the buildings had been destructed, with the exception of a part of the roof in front of the entrance of the mine and the engine house (without engines), and the entrance to the mine had been filled in (Fig. 1.4). The estate had been sold to a man who had returned from Angola. He did not want to permit any work on his land. He had had some drillings done for water for his plantations, but without success. The prospect of the water from the mine, which we mentioned, finally convinced him to give us the permission to open a hole in the entrance. G. KRUSAT entered through this hole with a diving breathing apparatus. He noticed that at least the beginning of the tunnel, which was supported by brickwalls, was intact, but that the mine was really filled with water. The reopening of the mine thus seemed

Fig. 1.5. Block diagram of the Guimarota mine. The beds dip southwards at an angle of 22°, the winding gallery is situated in the upper coal seam and a tunnel leads horizontally through the intercalated limestones into the lower seam, the so-called Fundflöz. The area that was mined from 1973 to 1982 is situated in the tectonically lowered block (right).

problematical, but technically not impossible.

It was more difficult to get the permission for the excavation from the landowner. After tough negotiations he gave in, in the prospect of the promised large amounts of water, completely free of charge, and a considerable rent for the buildings, which were unused at the time. However, he did not want to have any coal on his land, which later forced us to return the mined coal to the mine after it had been processed for the paleontological analysis.

We obtained the necessary permissions for the planned work from the Geological Survey of Portugal under the following three conditions: a local official has to check that the security measures in the mine are correctly followed; the recovered fossils have to be deposited in the museum of the Geological Survey in Lisbon after the scientific studies are finished; the results of the scientific work should be published in the publications of the Geological Survey (a condition that had been agreed on previously).

Now I was able to submit an application to the Deutsche Forschungsgmeinschaft, in agreement with the other members of staff of the institute. Only W. G. KÜHNE still dissociated himself from the project and told us "you are crazy!"

We had tried in vain to arouse the interest of another paleontological institution for a joint conduct of the project, and the proposal to the Institute of Mining of the Technische Universität Berlin, to use the planned mining campaign as a possibility to gain practical experience, did not find enough supporters in this institution. Therefore, the application for our project was submitted as a regular grant application to the Deutsche Forschungsgmeinschaft on February 17th, 1972, under the headword "Mammals in the Mesozoic". In the face of the considerable uncertainties that threatened the project, it was only granted for one year, and, after initial success, four times extended for another year. Then, the longest possible duration of funding by the Deutsche Forschungsgemeinschaft under the guidelines lined out in its constitution was exhausted. To keep up the excavation work, we applied for a special research project at the Freie Universität Berlin, which, under the title "paleontological analysis of non-marine depositional environments in terms of the evolution of small tetrapods, in particular mammals", had a broader scope, but was mainly restricted to the continuation of the work in the Guimarota mine. Thus, we could continue the excavation for five more years. The costs for the

Fig. 1.6. Partially collapsed tunnel before the reconstruction.

Fig. 1.7. Reconstruction of the framing walls of a collapsed tunnel in the spring of 1973.

field work in the Guimarota mine (salaries for the Portuguese employees, costs for material and equipment, travel costs) added up to a total of DM 857,550. Of this sum, the Deutsche Forschungsgemeinschaft supplied DM 482,950 and the Freie Universität Berlin DM 374,600.

We have only little information about the history of the brown coal mine of Guimarota. The mine had obviously been opened in the 1930ies, probably mainly to supply the cement factories of Leiria with fuel. The mine was situated approximately 2 km south of the center of the town of Leiria, on the road leading towards Fatima, near to a group of houses that bear the name Guimarota at the junction with the road towards Vidigal. The entrance of the mine was found on the slope of a hill in the west of the valley of the river Liz. Today, the former garden- and vine plantation-land is covered by buildings, it became the city district Guimarota. Only the name of the dead end street that formerly led to the shaft, "Rua das Minas" still reminds of the mine.

In the area of the mine, the beds dip southwards at an angle of 22° (Fig. 1.5). Their continuation is interrupted by several faults that show a north-south strike. Two coal seams, separated by eight meters of limestones, were mined. The former winding gallery was situated in the upper seam. This broad, oblique tunnel, which was also inclined at an angle of 22°, contained the two tracks for mine cars and a staircase for pedestrians. The first thirty meters of the tunnel were laid out in stonework. From the winding gallery, perpendicular tunnels branched off in regular intervals in the direction of the strike of the coal seam in both directions; it was in these tunnels that the coal was mined. From several of the perpendicular tunnels, a horizontal gallery led through the intercalated limestone into other tunnels in the lower seam, which were also arranged perpendicular to the dip of the beds. We knew that the coal that had yielded the mammalian remains in the first campaigns had come from the lower coal seam and was mined through the gallery branching off the fifth perpendicular tunnel in the east. This point had to be reached again. The distance from the entrance of the mine to the fifth perpendicular tunnel was 200 m, from there to the lower coal seam, it was an additional 60 m; at this point, one would be 80 m below the surface of the earth.

The reopening of a mine that had been closed for ten years and completely drowned is a very risky venture. It was thanks to the organizational abilities, the venturesome character and the powers of endurance of S. HENKEL, who took over the technical lead of the project, that our plans could be realized. HENKEL managed to take on the son of the former capataz (pithead foreman), GUILHERME FERNANDOS DOS SANTOS, as foreman for our work, as well as one of the carpenters of the mine and several miners who had worked in the mine before and thus had the necessary experience. Since it was impossible to employ these qualified workers on a monthly basis, we decided to work all year round, in contrast to our original intentions.

On August 1st, 1972, we started to clean out the entrance of the mine. Since the original roof was missing here, we built a new canopy. We started to pump the water that had accumulated in the mine out of the tunnels and into a reservoir on a higher level. The well preserved part of the winding gallery that had been built in brickwork led into a gallery that was supported by wooden frames, which were quite rotten (Fig. 1.6). Following the falling water level, we had to set new frames into place, one between every two old ones (Fig. 1.7). Thus, we needed large amounts of wood for the work in the mine. Therefore, S. HENKEL bought a small eucalyptus grove and had the trees cut down. A ventilator and air pipes were installed for the ventilation of the mine. Soon, we got to a stretch of tunnel of c. 30 m length that had collapsed, and above which a dangerous hollow had formed. The gallery had to be cleared out and rebuilt. On its roof, debris was filled in, to evenly distribute the weight of possible falling rocks. In the beginning of December 1972, when 150 m of the winding gallery had been cleared out and newly secured, another collapsed section blocked the way. S. HENKEL, who was the only one of us who was in Portugal at that time, was uncertain if the surmounting of this new obstacle was possible and acceptable, given its unknown extension. He decided to stop the work for the time being. In February of the following year, after we had examined the mine together, we decided, in the light of the work achieved so far, to take the risk and continue the reconstruction of the gallery. In April 1973, we reached the lower coal seam through the gallery that branched off from the fifth perpendicular tunnel. Tracks were laid down, a luckily preserved mine car was set on the tracks, a used electrical ship winch, which had been brought all the way from the Baltic sea, was set up as winding engine and electrical cables for the illumination and the electric jackhammer were laid down.

However, a big disappointment awaited us in the finally accessible part of the mine, which had taken us months of toilsome work to reach: all the coal had been completely cleared out here. The only thing to do was to "nibble" at the safety columns that had originally been left standing, and to mine coal from the margins of the tunnel. The examination of the coal soon proved to be successful. On Mai 14th, 1973, G. KRUSAT, who had supervised the begin of the mining work in the tunnel, reported the discovery of the first mammalian jaw, including the dentition, news that were duly celebrated in the institute.

However, in the long term, no effective mining was possible in this area. In the winter of 1972/1973, the miners cleared out a dead-end orientation tunnel which led, after crossing a fault line and through another tunnel in the overlaying limestones, to a still untouched area of the lower coal seam, in a different block that had been lowered in relation to the other layers along the fault. In this new area, called "Novas Galerias" by the miners, we could now freely mine coal for some time.

It showed that the mammalian remains were not evenly distributed in the sediment. There were obviously areas which were enriched in mammalian fossils (which means that we approximately found one fossil a day); these areas were termed "discovery bodies" by us. They could not be distinguished from other areas by any differences in the appearance of the coal. We developed several theories about the origin and distri-

Fig. 1.8. The pithead foreman GUILHERME FERNANDOS DOS SANTOS in the coal face, the height of which corresponds to the thickness of the mined coal seam.

Fig. 1.9. Miner ANTONIO SOARES mining coal using the electric jackhammer.

Fig. 1.10. The mine car in the perpendicular tunnel. On the right the air pipes for the ventilation.

were originally left standing.

The coal seam that contained the vertebrate remains (termed "Fundflöz") is only fifty to sixty centimeters thick; the coal could only be mined lying in the low coal face (Fig. 1.8). It made the work easier, however, that we used an electric jackhammer instead of a pickaxe (Fig. 1.9). To mine the different layers of coal separately was impossible, the complete thickness of the seam had to be removed as a whole. Thus, the different finds cannot be referred to any distinct layer within the coal seam. The mined area was noted down in the map of the mine each week, together with the date. The date of mining for each car load of coal was kept with the sediment in the further stages of its processing. Each day, one car load of coal was mined, which represents the equivalent of roughly one metric ton.

The loaded car was pushed along the horizontal section to the turntable in the winding gallery, where it was then pulled to the surface by the winch (Fig. 1.10). Agreed signs, given in the form of knocks on an iron wire, told the "engineer" in a traditional way when to operate the winch. The field telephone that we had installed originally, was completely corroded from the moist, acidic air in the mine after a short while and thus unusable. An iron bar that the car towed behind it was used as an emergency anchor in the event that

bution of such discovery bodies in the extension of the coal seam and accordingly followed different directions in the mining work. For some time, we even followed the dip of the seam, below the level of the tunnel. On the plan of the mine, our progress looked like the work of drunk miners, or like a mysterious place of worship. When we met faults in the west and east in the progress of our work, we also removed the pillars of coal that

Fig. 1.11. Portuguese helpers splitting and examining the coal in the hall, under the supervision of Dona ENCARNAÇÃO (left in the picture).

Fig. 1.12. Overview over the hall. On the left the binokular microscope for examining the finds and picking the residue resulting from screen washing; behind it, equipment for the HENKEL-process is visible. The entrance to the mine can be seen in the background; sediment is being processed to the right of it. The heap in the middle consists of newly mined coal; the tarp prevents the coal from drying out. On the right the table with the helpers. The tubs in front of the table receive the examined coal for further processing.

the cable of the winch broke. Unfortunately, we could not use this security measure when we brought the examined coal back into the mine, i. e. on the downhill journey of the loaded car. Once, a derailment of the car really resulted in significant damage.

The examined coal, broken up into small pieces, which we – as noted above – were not permitted to deposit on the campound of the mine, was first used to fill in unused tunnels close to the surface. Caused by chemical reactions, the coal heated up and this resulted in a local fire in the mine, which, thanks to the abundant water, was rapidly put out again. After this incident, we deposited the examined coal in deeper, still flooded tunnels.

In this context, it might be of interest to note that there were no poisonous or explosive gases in the Guimarota mine; dangerous was only the lack of oxygen.

The car with the mined coal was emptied in the roofed hall in front of the mine entrance. There, a long table had been set up, at which up to eight female Portuguese helpers split the coal with small hammers and knifes and examined it conscientiously for teeth, scales and bone remains, under the supervision of the wife of foreman, Dona ENCARNAÇÃO (Figs. 1.11 and 11.12). Fossils that were discovered on the split surfaces were encircled with chalk, and the chunks of coal were deposited on a tray. Each of the workers had her own color of chalk. The discovered objects were examined by a paleontologist at the end of each day, and specimens that were deemed to be worth keeping were roughly identified and packed orderly and securely. Since the excavation went on through the whole year, and we could only be in Portugal in the free time between university terms, Dona ENCARNAÇÃO later carried out this work. She accepted the additional, demanding responsibility with great enthusiasm, and soon gained a well-funded expertise on the faunal components of the coal of Guimarota.

The different colors of the chalk circles that marked every find enabled us to trace back every discovery to the respective helper. Thus, the attention and diligence of the helpers could be evaluated and bonus money could be paid. The amount paid to the helpers was made up of a basal salary and additional bonuses that were paid – graded according to the group of organism – for jaws of mammals and dentitions of other vertebrates.

With few exceptions, only small vertebrates

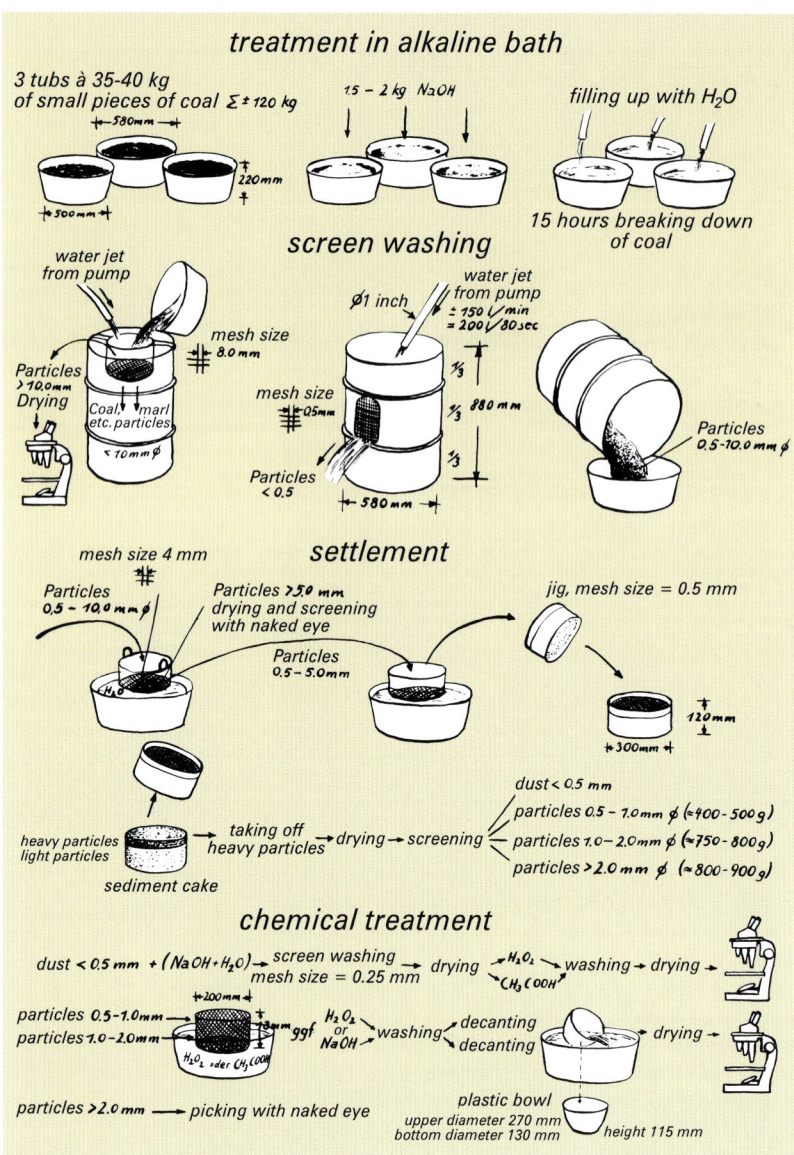

Fig. 1.13. Schematic overview of the coal processing using the HENKEL-process.

amount of water. If dried, the sediment crumbles, thus destroying the fossils contained within. Therefore, the chunks of coal that contained vertebrate fossils were packed solitarily in moist paper – the more valuable later also in foil – preliminary sorted, and shrink-wrapped in plastic bags. In the process of packing, each object was labeled with its discovery date, so that its provenance from within the horizontal extend of the coal seam can be determined with the help of the mining plans (see chapter on preparation).

Apart from the work described above, which allowed the discovery of more complete skeletons, a certain amount of coal was screen-washed each day, using the HENKEL-process mentioned above (Fig. 1.13). The aim was to determine the total amount of phosphate in the mined coal in order to obtain data on the percentile distribution of different faunal elements, as well as record changes in this data within the extension of the coal seam.

Each day, approximately 100 kg of coal were distributed in three plastic tubs which were then filled up with diluted potash lye and were left standing over night. On the following morning, the dissolved coal could be screen washed, using the HENKEL-process. The most important piece of equipment for this process is a (used) 200 l oil- or fuel barrel, in which a rectangular window is cut at half-height. A wire netting is welded into this opening, the screen size of which holds back the smallest objects expected; a size of 0.5 mm is suited for isolated teeth of Mesozoic mammals. Furthermore, an additional sieve is needed, which is put into the barrel. The size of this screen is meant to allow the passage of the largest expected objects; in Guimarota we used a sieve of 5 mm. The dissolved coal was poured into the additional sieve and washed through it with water. The material that accumulated in the sieve was emptied out and searched for possible larger vertebrate remains. The smaller fraction was washed with a jet of water until the water pouring out through the lateral opening remained clear. The water from the mine collected in the reservoir provided the necessary water pressure for this work. The concentrated sediment with a particle size between 0.5 and 5 mm, which remained in the barrel was then further processed by separating particles of different specific mass from each other – a procedure known from the processing of ore. For this process, we shook a certain amount of the concentrated sediment in a sieve of 0.5 mm under water by hand, and then emptied it out on a table. The heavier particles were now found on the top of the heap of turned sediment and could

were found in the coals of the Guimarota mine, either juvenile individuals or taxa that remained small as adults. The animals were usually represented by isolated bones, teeth, and scales (see chapter on the taphonomy). The fossils do not occur on the splitted plains, as is the case in lithographic limestones, but are only partially visible or even exposed in cross-section on the split surfaces of the poorly laminated sediment. A determination of the discovery is only possible if teeth or characteristic surface structures of bones are visible; otherwise, only the preparation of the specimen will reveal its identity.

The coal from Guimarota contains a high

be collected with a spoon. These parts, which also included the phosphate, were dried, fractioned and partially further processed, using other physical or chemical methods, for example a treatment with diluted acetic acid to dissolve calcareous particles.

The teenage son of the foreman, called GUILHERMECITO (little GUILHERME), picked the phosphate out of the concentrated sediment resulting daily from the process described above. Every day, he sorted teeth of mammals and "baby"-dinosaurs, tiny jaws of amphibians and lizards, and interesting bones into three separate cells; the remaining bone-, teeth- and scale-material was stored in small plastic boxes, also separately for each day. The approximately 6,000 "GUILHERMECITO-cells" are only now being analyzed in our institute. They provide invaluable statistical data for the taphonomic analysis of the lagerstatte, as well as data for biometrical investigations. Furthermore, as it was expected, they contain the rarest elements of the fauna, which were not sampled in the splitting of the coal.

For ten years, we worked in Guimarota in the way described above. Then the excavations had to be stopped. Apart from the temporal limits of the funding, other reasons also played a role in this decision: the general condition of the mine steadily worsened; the frames that we had built to secure the tunnels were rotting; the equipment became increasingly unreliable. It would have taken considerable amounts of money to continue the work. Since we furthermore did not find any new faunal elements in the last years of the excavations, we also assumed that the Late Jurassic fauna that had lived in this area was sampled in its complete diversity. In the end of 1982, the works were stopped, the mine was closed, and is now filled with water again.

The Guimarota mine has yielded an enormous wealth of fossil material. Approximately 10,000 isolated teeth, more than 1,000 jaws with dentitions in different states of preservation, c. 20 more or less compressed skulls and two almost complete skeletons of mammals alone have been recovered. Apart from that, uncounted remains of cartilaginous and bony fishes, of salamanders, turtles, lizards, crocodiles, pterosaurs, and dinosaurs are present. The preparation of this valuable material is far from being finished – as it is usual for paleontological excavations of this scale. Only the mammalian remains have all been prepared, as far as they had been identified in the field. The prepared material of other vertebrates should be representative enough by now to allow a scientific study of other vertebrate groups. Although the list of publications on the flora and fauna of Guimarota contains more than one hundred titles so far, the scientific work on the material will certainly take many more years to come.

Biographic information about deceased colleagues

HELMDACH, FRIEDRICH-FRANZ, 1935-1994. Studies at the Freie Universität Berlin, diploma (1966) on the geology of the Guimarota mine and doctoral dissertation (1968) on the ostracodes of the Guimarota mine under W. G. KÜHNE. From 1968 to 1971 scientific assistant, and then assistant professor at the Institute for Paleontology of the Freie Universität Berlin until 1978. Habilitation in 1975. After 1978 five years work at the University of Amman (Jordan). Afterwards commissions as a lecturer at the Institute of Geology of the Freie Universität Berlin.

HENKEL, SIEGFRIED, 1931-1984. Studies at the Freie Universität Berlin, diploma and doctoral dissertation (1960) on topics of the geology of the alps under M. RICHTER. 1959 supporting assistant of W.G. KÜHNE, 1963 scientific assistant to the newly founded professorship and later Institute of Paleontology at the Freie Universität Berlin. 1971 Appointment as professor. (Biography: KREBS 1985).

KRUSAT, GEORG, 1938-1998. Studies at the Freie Universität Berlin, diploma (1966) on the geology of Montsech in Spain under W. G. KÜHNE and doctoral dissertation (1974) on the docodonts of the Guimarota mine under G. HAHN. From 1966 to 1974 scientific assistant, since 1975 curator at the Institute of Paleontology of the Freie Universität Berlin.

KÜHNE, WALTER GEORG, 1911-1991. Studies in Berlin, later in Halle/Saale under JOHANNES WEIGELT. 1933 expelled, earned a living dealing with fossils. 1938 emigration to England, 1940-1944 internment, then collection work and scientific investigation of the therapsid *Oligokyphus* under D. M. S. WATSON in London. Doctoral dissertation on *Oligokyphus* 1949 (published 1956) at the University of Bonn. 1952 return to Germany. 1956 lecturing commission for paleontology at the Geological-Paleontological Institute of the Freie Universität Berlin. Habilitation 1958. 1963 appointment as professor at the at the newly founded professorship and later Institute of Paleontology at the Freie Universität Berlin. Retired in 1976. (Biographies: KOHRING & SCHLÜTER 1991, 1997, KREBS 1991, SCHLÜTER 1981).

References

HELMDACH, F. F. (1966): Stratigraphie und Tektonik der Kohlengrube Guimarota bei Leiria (Mittel-Portugal) und ihrer Umgebung. – Unpublished Diploma thesis, Freie Universität Berlin. – 75 pp., Berlin.

HENKEL, S. (1966): Methoden zur Prospektion und Gewinnung kleiner Wirbeltierfossilien. – Neues Jahrbuch für Geologie und Paläontologie, Monatshefte **1966**: 178-184.

KOHRING, R. & SCHLÜTER, T. (1991): In memoriam WALTER GEORG KÜHNE (26.2.1911-16.3.1991). – Berliner geowissenschaftliche Abhandlungen A **134**: 3-8.

– (1997): Einleitung zu: Paläontologische Essays 1943-1990 von WALTER GEORG KÜHNE. – Documenta naturae **113**: 3-24.

KREBS, B. (1985): SIEGFRIED HENKEL 1931-1984. – Berliner geowissenschaftliche Abhandlungen A **60**: 1-4.

KREBS, E. (1991): WALTER GEORG KÜHNE 26.2.1911-16.3.1991. – Paläontologie aktuell **24**: 18-22.

KÜHNE, W. G. (1961a): A mammalian fauna from the Kimmeridgian of Portugal. – Nature **192**: 274-275.

– (1961b): Eine Mammaliafauna aus dem Kimmeridge Portugals. Vorläufiger Bericht. – Neues Jahrbuch für Geologie und Paläontologie, Monatshefte **1961**: 374-381.

– (1961c): Une faune de mammifères lusitaniens (rapport provisoire). – Comunicaçoes dos Serviços geológicos de Portugal **45**: 211-221.

– (1968): History of discovery, report on the work performed, procedure, technique and generalities. Contribuição para a Fauna do Kimeridgiano da Mina de Lignito Guimarota (Leiria, Portugal) I Parte, I. – Memórias dos Serviços geológicos de Portugal, (nova Sér.) **14**: 7-20.

LAPPARENT, A.F. DE & ZBYSZEWSKI, G. (1957): Les Dinosauriens du Portugal. – Memórias dos Serviços geológicos de Portugal, (nova Sér.) **2**: 1-63.

SCHLÜTER, T. (1981): WALTER GEORG KÜHNE 70 Jahre alt – eine biographische Skizze. – Berliner geowissenschaftliche Abhandlungen A **32**: 1-17.

Geological setting and dating of the Guimarota-beds

Michael E. Schudack

Geological setting

The vertebrate bearing beds of the Guimarota mine are only locally exposed and do not contain any marine index fossils; therefore their correlation with other layers of the closer surroundings is difficult. Their exact age has been a matter of debate up to the present day. Within the regional geological framework, the fossiliferous beds ("Guimarota-beds") can be assigned to the Alcobaça Formation, which is usually correlated with the Abadia Formation, which can be dated with the help of ammonites (Fig. 2.1).

In the area of the mine, a total thickness of approximately 20 m of the beds has been exposed (Helmdach 1971). Within the layers, a lower coal seam ("Fundflöz") is separated from an upper one ("Ruafolge") by an approximately 5 m thick layer of limestone (Fig. 2.2). The seams consist of alternating layers of marls, which are sometimes rich in fragmented bivalve shells, and marly coals (Fig. 2.3). Pure coal seams are rare and of a thickness of only a few centimeters. Further information on the sedimentology can be found in Helmdach (1971) and Henkel & Krusat (1980).

Invertebrates are mainly represented by fresh and brackish water bivalves and snails (Aberhan et al. 2000*) and by a diverse ostracode fauna. In a first paper on the gastropods by Bandel (1991), two new species, *Ptychostylus guimarotensis* and *Melampoides jurassicus*, were described, which probably lived in brackish water and the spray zone. The ostracodes from the Guimarota mine are of outstanding importance for both the dating (Kimmeridgian), as well as the interpretation of the paleoecology of the fossiliferous beds and

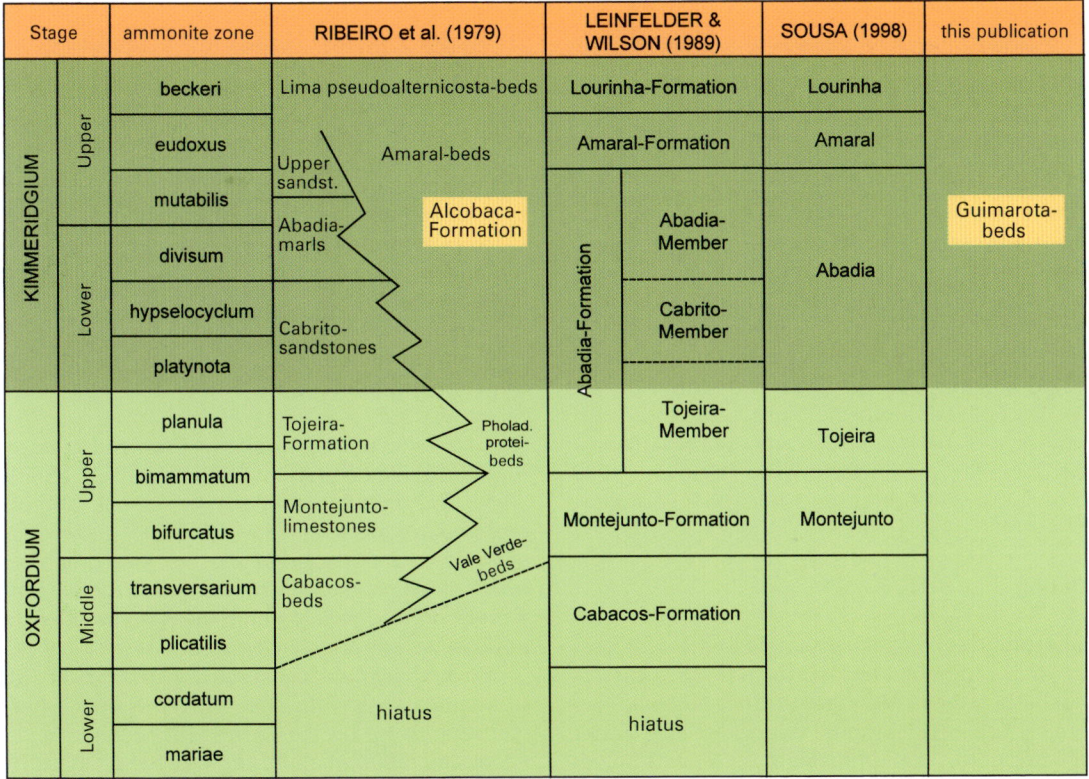

Fig. 2.1. Correlation of the Guimarota beds (this work) and the lithostratigraphic units of the Oxfordian and Kimmeridgian in the region of Montejuntao – Torres Vedras – Leiria, based on Ribeiro et al. (1979), Leinfelder & Wilson (1989) and Sousa (1998). Vertical temporal axis not to scale.

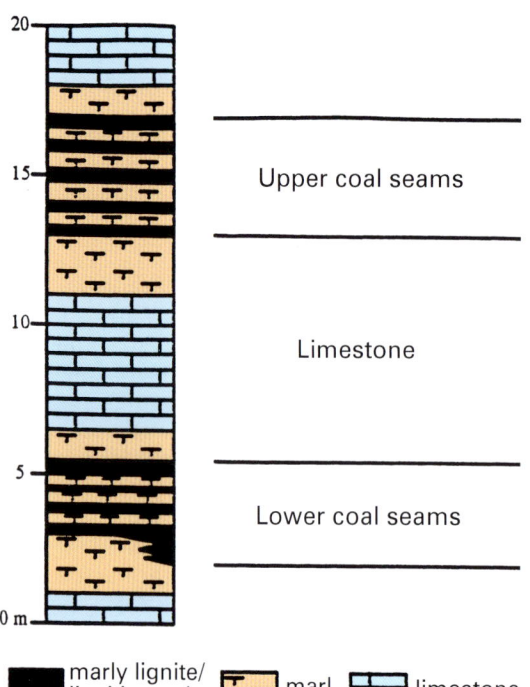

Fig. 2.2. Strongly simplified schematic section of the layers exposed in the Guimarota mine, showing the position of the analyzed samples. From SCHUDACK (1993a), based on HELMDACH (1971).

they will be dealt with in a separate chapter (SCHUDACK 2000*).

The sparse macroflora from the Guimarota mine was studied by BRAUCKMANN (1978), who attempted to use it for biostratigraphical and paleoecological interpretations (restudied by MOHR & SCHULTKA 2000*). Microscopic plant remains are much more common and include gyrogonites (calcified female oogonia) of charophytes (a kind of algae) as well as spores and pollen. The charophytes were already used in an early phase of the history of research on the Guimarota mine to date the fossiliferous beds. According to the determinations of MÄDLER in HELMDACH (1968, 1971), they indicate a Kimmeridgian age. Spores and pollen were studied later (VAN ERVE & MOHR 1988, MOHR 1989, MOHR & SCHMIDT 1988). In contrast to the results obtained from the ostracodes and charophytes, the palynomorphs apparently indicated an Oxfordian age for the layers. These contradictory dates will be discussed in some detail below.

Dating

Charophytes: The first attempts to date the locality were carried out by HELMDACH (1968, 1971), on the basis of the gyrogonites of charophytes, which are very abundant in some layers. He listed (based on taxonomic determinations by KARL MÄDLER, Hannover) the species *Porochara westerbeckensis* and *Porochara raskyae* and proposed a "Kimmeridgian- to Lower Portlandian"-age on this basis. A restudy of the charophyte flora (SCHUDACK 1993a) furthermore resulted in the identification of *Porochara fusca* var. *minor* and *Mesochara* sp., which is consistent with an Upper Oxfordian to Kimmeridgian age. However, such forms should be only used with greatest caution for exact dating, since it is meanwhile known that many of them have long stratigraphic ranges. Unfortunately, no representatives of the stratigraphically more reliable family Clavatoraceae were found in the Guimarota beds so far. In all cases, in which taxa known from the Guimarota mine have also been found in the better dated coastal sections of Portugal (GRAMBAST-FESSARD & RAMALHO 1985), they are restricted to the Kimmeridgian and are still absent in the Upper Oxfordian (SCHUDACK 1993a).

Ostracodes: Several different species of ostracodes enabled HELMDACH (1971) to determine the age more precisely. He stated that *Cetacella inermis* allowed a dating of the beds as "Kimmeridge", and *Oertliana* (now *Dicrorygma*) *kimmeridgensis* even more precisely as "Lower Kimmeridge". However, the dating on the basis of the species of *Dicrorygma*, which was washed in from marine waters, cannot be supported, since HELMDACH (1971) did obviously not take the nowadays usual distinction between the "Kimmeridge" (sensu anglico, used only locally in England) and the Kimmeridgium sensu stricto (sensu gallico, used internationally) into consideration. After all, the same species was reported by OERTLI (1957) from the middle "Kimmeridgien inférieur" of the Paris basin (= boundary zone of the Lower/Upper Kimmeridgian in the international use), by KILENYI (1965, 1969) in the *mutabilis*-zone of southern England, which is precisely dated on the basis of ammonites (= lowermost Upper Kimmeridgian in the international use, see Fig. 2.1) as well as by SCHMIDT (1955) – in open nomenclature – in the German Middle Kimmeridge (= lowermost Upper Kimmeridgian in the international use). Meanwhile, *Dicrorygma kimmeridgensis* has furthermore been synonymised with the species *D. reticulata* (see SCHUDACK 2000*), the stratigraphical range of which extends into the Tithonian, so that a more precise dating within the Kimmeridgian is impossible anyway.

Several authors accepted the referral of the Guimarota beds into the "Lower Kimmeridge" (sensu anglico) by HELMDACH (1971), without the necessary differentiation, and regarded it as equal to the Lower Kimmeridgian in the international use (sensu gallico; e.g. BRAUCKMANN 1978, SCHMIDT 1985, 1986, MOHR & SCHMIDT 1988). A revision of the Late Jurassic ostracodes from several sections of central Portugal, which unfortunately remained unfinished due to his sudden death, led HELMDACH to the assumption that the ostracode fauna from the Guimarota mine represents an ostracode association that is widely distributed in central Portugal in the Lower Kimmeridgian (sensu gallico). Therefore, SCHUDACK (1993a) also assumed a Lower Kimmeridgian age (sensu gallico) for the Guimarota beds.

In a conversation with the author, HELMDACH stated that he assumed a higher Lower Kimmeridgian age (*divisum*-zone), or possibly a lower Upper Kimmeridgian age (*mutabilis*-zone) for the Guimarota beds. This is in general accordance with the lithostratigraphical correlations published so far (RIBEIRO et al. 1979, LEINFELDER & WILSON 1989, see Fig. 2.1) and also fits well the current state of the art in terms of the stratigraphic ranges of the ostracode species that are present in Guimarota (SCHUDACK et al. 1998).

The often stated objection that the presence of ostracodes is too much dependant on the facies for them to be of value for a precise biostratigraphy is erroneous. The species of the nonmarine genera *Theriosynoecum*, *Timiriasevia* and *Cetacella*, for example, which also occur in Guimarota, are found in the Kimmeridgian of Europe, western North America (SCHUDACK 1995, 1996a, 1999, SCHUDACK et al. 1998) and China (LI 1983, WU et al. 1983). Several reports have been published on the easy and rapid distribution of the resistant eggs of nonmarine ostracodes as part of the aireal plankton (HELMDACH 1979, WHATLEY 1990, SCHUDACK 1999).

Macroflora: Because of the biostratigraphic uncertainties, attempts were made early to date the Guimarota beds with the sparse and usually poorly preserved macroflora. BRAUCKMANN (1978) stated that a precise date could not be established on the basis of the macroflora, but that the phylogenetic relationships of the flora would not prevcent the Guimarota beds to be "older than Kimmeridgian". This statement was repeatedly restated by later authors in a way that indicated that the macroflora of the Guimarota mine suggests an Oxfordian age, or even proved this dating more or less conclusively (VAN ERVE & MOHR 1988, MOHR 1989).

Microflora: Pollen and spores from the Guimarota mine are certainly of greater stratigraphic significance. On the basis of the microflora, VAN ERVE & MOHR (1988) concluded that the Guimarota beds were deposited in the Oxfordian. This statement was disputed by SCHUDACK (1993a), mainly because of new information on the stratigraphical range of the taxa used for the dating. Furthermore, the ecological peculiarities of the pollen and spores association were not given due considerations.

Fig. 2.3. The lower coal seam ("Fundflöz") exposed in the coal face. It is a marly coal, with small intercalated layers of shiny coal. The numerous white dots are mollusk shells. Scale in decimeters.

Relative dating on the basis of correlations with surrounding areas

According to HELMDACH (1971) the Guimarota beds, being part of the Alcobaça Formation, can be correlated with the marine Abadia Formation, which is exposed south of Guimarota (Lower Kimmeridgian sensu anglico according to RUGET-PERROT (1961), dated by ammonites). RIBEIRO et al. (1979) referred the Alcobaça Formation to the Middle Kimmeridgian (sensu gallico). Recently, the Abadia Formation (see Fig. 2.1) has been restricted to the lower and middle parts of the Kimmeridgian (sensu gallico; LEINFELDER & WILSON 1989).

Based on comparisons of the floras, BRAUCKMANN (1978) noted the possibility that the Guimarota beds could be correlated with the Oxfordian Cabacos layers. In a paper on the palynoflora of the area around Porto de Mós, which is very

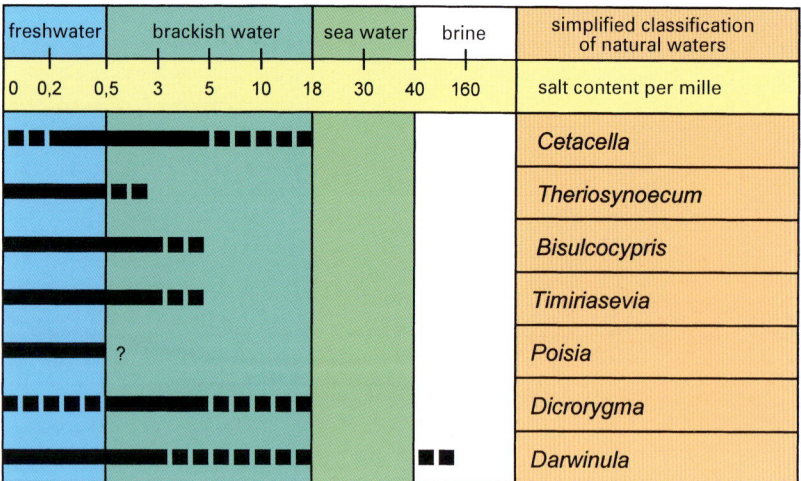

Fig. 2.4. Salinity tolerance ranges of the ostracode genera found in the Guimarota Formation so far. Modified from WEISS (1995) and SCHUDACK et al. (1998); see these papers for detailed sources of information.

similar to that of the Guimarota mine, MOHR & SCHMIDT (1988) reached a similar conclusion. Since MOHR & SCHMIDT (1988) found the dinoflagellate cyst *Sentusidium rioultii* in the overlying beds at Porto de Mós, which at that time was assumed to be typical for the Oxfordian, they transferred this date to the Guimarota beds. However, *Sentusidium rioultii* has been found since in younger layers at several localities (SCHUDACK 1993a).

All recent superregional correlations (Fig. 2.1) place the Alcobaça Formation and the Abadia layers (or -member) in the Kimmeridgian, which can thus be regarded as being certain.

Age of the Guimarota beds

The placement of the Guimarota beds within the Oxfordian, as suggested several times on the basis of paleobotanical evidence (micro- and macroflora), is not convincing. Unfortunately, vertebrates and macro-invertebrates cannot be used for a precise dating. Only the ostracodes have provided conclusive evidence so far: according to this evidence, the Guimarota beds are of Kimmeridgian age (sensu gallico; c. 151 to 154 million years according to GRADSTEIN & OGG 1996), which is also supported by the charophytes. However, a more precise dating of the layers within the Kimmeridgian is not possible at the moment, in contrast with earlier ideas (SCHUDACK 1993a).

Paleoecology

HELMDACH (1971) assumed that the Guimarota beds were deposited in a shallow and relatively small lagoon with fresh water inlets. The almost complete lack of clastic sediments (e.g. sandstones) indicates a hinterland with low relief. The salinity level of the water in the lagoon was obviously subject to variation, as indicated by the fact that, apart from the majority of organisms, which lived in fresh and brackish waters, both micro- and macrofauna contain marine forms, such as the marine crocodile *Machimosaurus hugii* v. MEYER, or representatives of the ostracode genus *Dicrorygma*, as well as a number of foraminifera (HELMDACH 1971). The dominant fresh water aspect was postulated by HELMDACH (1971) on the basis of the presence of the ostracode genera *Darwinula*, *Bisulcocypris* and *Theriosynoecum* and the great abundance of charophytes. However, whereas this interpretation can be confirmed for the ostracodes, the charophytes must be interpreted differently, based on recent research (SCHUDACK 1993a).

A new evaluation of the ecological requirements of the ostracodes from the Guimarota beds in terms of their salinity tolerances does not result in a uniform picture (Fig. 2.4), which can, however, be explained by the provenance of the examined material from different layers from within these varied intercalated beds (Fig. 2.2). The assemblage is strikingly dominated by genera that lived in fresh to slightly brackish waters, whereas no strictly marine genera have been identified so far.

HELMDACH's (1971) assumption that the massed occurrences of charophytes indicate fresh water conditions must be disputed, based on new insights, since not all taxa exhibit the same requirements in terms of salinity (SCHUDACK 1993a). Especially the genus *Porochara*, which is almost the sole representative found in the Guimarota beds, though present in fresh water environments, is frequently found in deposits that were laid down in brackish water. It is even more or less typical for such deposits, according to an analysis of the salinity requirements of the most important Late Jurassic/Early Cretaceous representatives of this genus (SCHUDACK 1993b, 1996a). Thus, with this high tolerance of different levels of salinity, it fits well with the picture of a lagoon with varying fresh- and brackish water conditions.

A further aspect is the massed occurrence of gyrogonites in the light of the absence of calcified vegetative stems of charophytes in the marly coals of Guimarota. If it is assumed that the gyrogonites were embedded at the same locality where the charophytes grew, this imbalance of the numbers is also typical for deposition under brackish conditions in the Late Jurassic and Early Cretaceous, since most of the typical fresh-water charophytes (clavatoraceans), which have strongly calcified stems, do not occur in these environments, due to the high salinity (SCHUDACK 1993b). Thus, vegetative parts (fragments of stems) are generally rare in brackish water deposits, with the exception of the genus *Echinochara*, which, however, is absent in Guimarota.

A detailed statistical analysis of more than 1,000 gyrogonites did not reveal any traces of mechanical flow sorting, which would have been expected if the fossils had been transported over longer distances (SCHUDACK 1993a). The statistical distribution of the populations equals those of known, certainly autochthonous (embedded at the locality where they lived) Recent populations. Thus, the charophytes sometimes formed extensive lawns in a brackish lagoon in Guimarota, and their oogonians have not been washed in from fresh-water environments.

References

ABERHAN, M., RIEDEL, F. & GLOY, U. (2000*): The mollusk fauna from the Guimarota mine. – In: MARTIN, T. & KREBS, B. [eds.] Guimarota – a Jurassic ecosystem: 37-40, München (Verlag Dr. F. Pfeil).

BANDEL, K. (1991): Gastropods from the brackish and fresh water of the Jurassic – Cretaceous transition (a systematic reevaluation). – Berliner geowissenschaftliche Abhandlungen A **134**: 9-55.

BRAUCKMANN, C. (1978): Beitrag zur Flora der Grube Guimarota (Ober-Jura; Mittel-Portugal). – Geologica et Palaeontologica **12**: 213-222.

GRADSTEIN, F. M. & OGG, J. (1996): A Phanerozoic time scale. – Episodes **19**: 3-4.

GRAMBAST-FESSARD, N. & RAMALHO, M. (1985): Charophytes du Jurassique supérieur du Portugal. – Revue de Micropaléontologie **28**: 58-66.

HELMDACH, F. F. (1968): Oberjurassische Süß- und Brackwasserostracoden der Kohlengrube Guimarota bei Leiria (Mittelportugal). – Unpublished PhD-thesis, Freie Universität Berlin. – 92 pp., Berlin.

– (1971): Stratigraphy and ostracode-fauna from the coal mine Guimarota (Upper Jurassic). – Memórias dos Serviços Geológicos de Portugal, N.S. **17**: 43-88.

– (1979): Möglichkeiten der Verbreitung nichtmariner Ostrakoden-populationen und deren Auswirkungen auf die Phylogenie und Stratigraphie. – Neues Jahrbuch für Geologie und Paläontologie, Monatshefte **1979**: 378-384.

HENKEL, S. & KRUSAT, G. (1980): Die Fossil-Lagerstätte in der Kohlengrube Guimarota (Portugal) und der erste Fund eines Docodontiden-Skelettes. – Berliner geowissenschaftliche Abhandlungen A **20**: 209-216.

KILENYI, T. I. (1965): *Oertliana*, a new ostracode genus from the Upper Jurassic of North-West Europe. – Palaeontology **8**: 570-576.

– (1969): The Ostracoda of the Dorset Kimmeridge Clay. – Palaeontology **12**: 112-160.

LEINFELDER, R. R. & WILSON, R. C. L. (1989): Seismic and sedimentologic features of Oxfordian-Kimmeridgian synrift sediments on the eastern margin of the Lusitanian Basin. – Geologische Rundschau **78**: 81-104.

LI, Y. (1983): On the non-marine Jurassic-Cretaceous boundary in Sichuan Basin by Ostracodes. (in Chinese with english summary). – Bulletin of the Chengdu Institute of Geological and Mineral Resources, Chinese Academy of Geological Sciences **4**: 78-89.

MOHR, B. A. R. (1989): New palynological information on the age and environment of Late Jurassic and Early Cretaceous vertebrate localities of the Iberian Peninsula (eastern Spain and Portugal). – Berliner geowissenschaftliche Abhandlungen, A **106**: 291-301.

MOHR, B. A. R. & SCHMIDT, D. (1988): The Oxfordian/Kimmeridgian boundary in the region of Porto de Mós (Central Portugal): stratigraphy, facies and palynology. – Neues Jahrbuch für Geologie und Paläontologie, Abhandlungen **176**: 245-267.

MOHR, B. A. R. & SCHULTKA, S. (2000*): The flora of the Guimarota mine. – In: MARTIN, T. & KREBS, B. [eds.] Guimarota – a Jurassic ecosystem: 27-32, München (Verlag Dr. F. Pfeil).

OERTLI, H. J. (1957): Ostracodes du Jurassique supérieur du Bassin de Paris (Sondage Vernon 1). – Revue de l'Institut Francais du Petrole **12**: 647-695.

RIBEIRO, A., ANTUNES, M. T., FERREIRA, M. P., ROCHA, R. B., SOAREZ, A. F., ZBYSZEWSKI, G., MOITINHO DE ALMEIDA, F., DE CARVALHO, E. & MONTEIRO, J. H. (1979): Introduction à la géologie générale du Portugal. – Serviços Geológicos de Portugal: 1-114, Lisbon.

RUGET-PERROT, C. (1961): Etudes stratigraphiques sur le Dogger et le Malm inférieur du Portugal au Nord du Tage. Bajocien, Bathonien, Callovien, Lusitanien. – Memórias dos Serviços Geológicos de Portugal, N.S. **7**: 1-197.

SCHMIDT, D. (1985): Faziesausbildung und Diapirismus im Oberjura von Mittel-Portugal am Beispiel des Diapirs von Porto de Mós. – Berliner geowissenschaftliche Abhandlungen A **60**: 49-89.

– (1986): Petrographische und biofazielle Untersuchungen an oberjurassischen Deckschichten des Diapirs von Porto de Mós (Mittelportugal). – Berliner geowissenschaftliche Abhandlungen A **77**: 1-211.

SCHMIDT, G. (1955): Stratigraphie und Mikrofauna des mittleren Malm im nordwestdeutschen Bergland. – Abhandlungen der senckenbergischen naturforschenden Gesellschaft **491**: 1-76.

SCHUDACK, M. (1993a): Charophyten aus dem Kimmeridgium der Kohlengrube Guimarota (Portugal). Mit einer eingehenden Diskussion zur Datierung der Fundstelle. – Berliner geowissenschaftliche Abhandlungen E **9**: 211-231.

– (1993b): Die Charophyten aus Oberjura und Unterkreide von Westeuropa. Mit einer phylogenetischen Analyse der Gesamtgruppe. – Berliner geowissenschaftliche Abhandlungen E **8**: 1-209.

SCHUDACK, M. (1995): Neue mikropaläontologische Beiträge (Ostracoda, Charophyta) zum Morrison-Ökosystem (Oberjura des Western Interior, USA). – Berliner geowissenschaftliche Abhandlungen E **16**: 389-407.

– (1996a): Die Charophyten des Niedersächsischen Beckens (Oberjura – Berriasium): Lokalzonierung, überregionale Korrelation und Palökologie. – Neues Jahrbuch für Geologie und Paläontologie, Abhandlungen **200**: 27-52.

– (1996b): Ostracode and charophyte biogeography in the continental Upper Jurassic of Europe and North America as influenced by plate tectonics and paleoclimate. – Museum of Northern Arizona Bulletin **60**: 333-342.

– (1999): Ostracoda (marine/nonmarine) and paleoclimate history in the late Jurassic of Central Europe and North America. – Marine Micropaleontology **37**: 273-288.

– (2000*): Ostracodes and charophytes from the Guimarota beds. – In: MARTIN, T. & KREBS, B. [eds.] Guimarota – a Jurassic ecosystem: 33-36, München (Verlag Dr. F. Pfeil).

SCHUDACK, M., TURNER, C. E. & PETERSON, F. (1998): Biostratigraphy, paleoecology, and biogeography of charophytes and ostracodes from the Upper Jurassic Morrison Formation, Western Interior, U.S.A. – Modern Geology **22**: 379-414.

SOUSA, L. (1998): Upper Jurassic (Upper Oxfordian – Tithonian) palynostratigraphy from the Lusitanian Basin (Portugal). – Memorias da Academia das Ciencias de Lisboa, Classe de Sciencias **37**: 49-77.

VAN ERVE, A. & MOHR, B. A. R. (1988): Palynological investigation of the Late Jurassic microflora from the vertebrate locality Guimarota coal mine (Leiria, Central Portugal). – Neues Jahrbuch für Geologie und Paläontologie, Monatshefte **1988**: 246-262.

WHATLEY, R. (1990): The relationship between extrinsic and intrinsic events in the evolution of Mesozoic non-marine Ostracoda. – In: KAUFMANN, E. G. & WALLISER, O. H., [eds.] Extinction Events in Earth History: Lecture Notes in Earth Sciences **30**: 253-263.

WEISS, M. (1995): Stratigraphie und Microfauna im Kimmeridge SE-Niedersachsens unter besonderer Berücksichtigung der Ostracoden. – Clausthaler geowissenschaftliche Dissertationen **48**: 1-274.

WU, Q., YANG, W. & HU, C. (1983): Upper Jurassic Ostracoda from Jishan of Anhui Province. (in Chinese with English summary). – Acta Palaeontologica Sinica **22**: 651-662.

The flora of the Guimarota mine

BARBARA MOHR & STEPHAN SCHULTKA

Unfortunately, the Jurassic flora of the Guimarota mine is only partially known. Only few macroscopic remains, mainly stem remains of horsetails, benettites (relatives of palm ferns) and conifers, were recovered from the rock layers of the former coal mine. This apparent poverty in terms of plant diversity can be explained by the processes of preservation and diagenesis, and also by the methods of sampling, which were focused on the discovery of vertebrate remains. However, PAIS (1998) also noted the low plant diversity in most Jurassic plant localities in Portugal. Thus, the microflora – especially the spores and pollen, which are randomly distributed in the sediment – plays an important role in the reconstruction of the vegetation, although it is also only moderately diverse, representing some 30 taxa (VAN ERVE & MOHR 1988).

Fig. 3.1. Sediment infilling of the stem of *Equisetites lusitanicum*, a common horsetail plant from the Late Jurassic of the Iberian Peninsula. Length 2.5 cm.

Horsetails

One of the dominant elements of the macroflora, the horsetail *Equisetites*, was described in detail by BRAUCKMANN (1978) under the specific name *E. lusitanicum*. These otherwise almost featureless fossils are well identifiable on the basis of the characteristic rock infillings of the core of their stem, which always exhibits well developed longitudinal channels (Fig. 3.1). Leaves and impressions of the usually smooth outer surface of the stem are generally rarer and are unknown from the material from Guimarota. Remains of these plants, which lived in shallow waters close to the shores, are often found in the layers surrounding the coal seams, but also in the seams themselves. In the latter sediment, however, the horsetails are only preserved as impressions of the cuticles, which are characterized by their well-developed pattern of rectangular rows of cells. In contrast, spores of equisetes (*Calamospora*) are rare, probably due to their thin walls. As it is the case in other Late Jurassic localities (SCHULTKA 1991), the equisetes probably formed monotypic groves.

Club mosses and ferns

Although club mosses and ferns are unknown in the macroflora from Guimarota so far, they constitute an important part of the pollen- and sporeflora. Whereas the lycophytes (club mosses) are mainly represented by *Lycopodium (Retitriletes)*, which forms 1-2 % of the association, fern spores account for approximately 36 % of the total spectrum, and, together with the conifers, represent the most important palynological unit. Their diversity is quite high, with approximately a dozen taxa being distinguishable. Spores of the families Osmundaceae, Gleicheniaceae *(Gleicheniidites)*, Cyatheaceae *(Deltoidospora*, among others), and Schizaeaceae *(Ischyosporites)* are present. *Ischyosporites marburgensis* (Fig. 3.2) is almost identical with spores found in situ of the fern *Klukia exilis* (VAN KONIJNENBURG-VAN CITTERT 1981), a genus of schizaeceans which is abundant in Jurassic strata. Today, this group of ferns lives on mineral-poor soils in periodically dry to semi-swampy environments (KRAMER 1990).

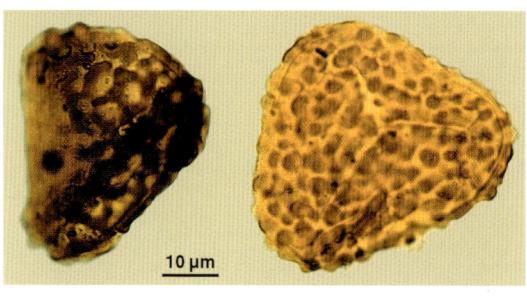

Fig. 3.2. *Ischyosporites marburgensis* DE JERSEY 1963 (left) and *Leptolepidites major* COUPER 1958 (right), fern spores from the coaly beds of the Guimarota mine. *Ischyosporites* can be referred to the family Schizaeaceae, which is still found today in subtropical to tropical regions.

Members of the family Matoniacea are probably also present, represented by forms such as *Deltoidospora mesozoica*, which typically exhibits a thickened laesura, slight thickenings in the interradial wall areas and a slightly reticulate surface of the perispor (VAN KONIJNENBURG-VAN CITTERT 1993). These spores are similar to those of *Phlebopteris muensteri* (VAN KONIJNENBURG-VAN CITTERT & KURMANN 1994), a species that has a wide distribution from Greenland to northern Africa in the higher Mesophytic (HIRMER & HÖRHAMMER 1936).

Pollen of pteridosperms were found in layers of similar age at the Serra de Porto de Mós, close to the Guimarota mine (MOHR & SCHMIDT 1988). *Vitreisporites* is not common there (c. 5 %), but nevertheless proves the presence of Caytoniales (PEDERSEN & FRIIS 1986). *Eucommiidites*-pollen represent a now extinct group of anthophytes (PEDERSEN et al. 1989), a clade that also includes the gnetales and the angiospermes (flowering plants). *Eucommiidites* was produced in the pollen-bearing organ *Eucommiitheca*, which has only recently been described from Early Cretaceous sediments of Portugal (FRIIS & PEDERSEN 1996).

Fig. 3.3. Fossil charcoal (fusite) of a strongly fragmented wood of a cycad, with a transfusion cell near the left margin (scanning electron-microscopic photograph); scale bar = 10 µm.

Cycads (palm ferns)

Cycad-like plants (palm ferns) are known from the Guimarota mine from both macroscopic and microscopic remains. Thus, transfusion cells (determination by H. SÜSS, Berlin; Fig. 3.3) could be detected in a piece of fusite (fossil charcoal). The transfusion tissue accompanies the tracheids and arranges the transmission of materials to the mesophyll. This kind of cells is present in the leaves of most gymnosperms (plants whose seeds are not enclosed in an ovary), but only in cycads also occurs in the pith of the stem (GREGUSS 1968). Since the specimen examined does not represent a fusite derived from leaves (REMY 1954), this find is a macroscopic proof for the existence of cycads in the flora from Guimarota.

The Cycadophytina are also represented by the pollen genus *Cycadopites*. This genus is relatively rare, forming only 4 % of the total association, but it is represented by several different species, which indicate both true cycads, as well as benettites; the latter group is restricted to the Mesozoic.

A partial leaf of *Otozamites mundae* (Fig. 3.4) probably comes from the layers overlying the lower coal seam, which was mainly mined. The genus *Otozamites* is referable to the benettites, based on the structure of the stomata. In its general appearance, *Otozamites* was probably similar to the Recent palm fern *Cycas revoluta*. The elongate leaflets of the leaf exhibit a fine fanlike veining, which turns into an apparently parallel veining in the anterior part of the leaflet. The point-like attachment of the leaflets on the main axis of the leaf, with an ear-like, anadrome bulge of the basis of the leaflet, is typical for the genus. The bulge is only weakly developed in the species *Otozamites mundae*. The genus is well known from the "deltaic series" (= Ravenscar Formation) of Yorkshire (Northeast England), where it represents the flora which played no or only a minor role in the formation of the coal seams. Thus, in the Guimarota ecosystem, the genus provides evidence for the flora of the hinterland, which grew on better drained, more permeable soils.

Conifers

Whereas the horsetails are typical for wet locations, conifers are the dominant plants of the dry and humid locations in areas both near to coal swamps, as well as those further away from them. They are by far the most abundant group of plants in the Middle to Late Jurassic of central and southern Europe. Apart from their original abundance, their dominance in fossil assemblages is enhanced by taphonomic processes, since their dense woods and resin-rich, needle-like leaves can be transported over long distances. Furthermore, they are resistant against biological decay processes for a long time. Consequently, they represent the dominant floral element in both the macro- and microflora (more than 40 % of the total association) in the Guimarota mine. Morphologically, *Brachyphyllum* (Fig. 3.5) is a genus with thick and large leaves, whereas a genus with smaller and more slender leaves is called *Pagiophyllum*. Both genera represent collective taxa that only stand for two principal morphologies. Many of these leaves probably represent members of the araucariaceans (HARRIS 1979), others are members of the cheirolepidiaceans. Accord-

ing to PAIS (1998) only cheirolepidiaceans are found in macrofloras from the Late Jurassic of Portugal. However, the fact that almost 20 % of the pollen association represent several different species of the genus *Callialasporites*, which is usually regarded as a type of araucarioid pollen (Fig. 3.6), seems to contradict this. *Callialasporites* has been isolated from a cone of the conifer *Brachyphyllum mamillare* from the Jurassic of Yorkshire (VAN KONIJNENBURG-VAN CITTERT 1971). However, the cones of *Hirmeriella (Brachyphyllum) crucis* and *Pagiophyllum kurri* contain the pollen genus *Classopollis* and thus represent members of the Cheirolepidiaceae. *Classopollis* and *Corollina* are also common in the Guimarota mine and make up 20 % of the pollen association. The cuticles of araucarians like cheirolepidaceans exhibit typical xeromorph (adapted for dry conditions) characters, such as sunken stomata, air holes that are protected by papillae, and strongly cutinised cell walls.

Taxodiaceans are probably represented by *Cerebropollenites macroverrucosus* in the pollen association (Fig. 3.6). This pollen is very similar to those isolated from *Masculostrobus*, a genus that VAN KONIJNENBURG-VAN CITTERT et al. (1989) tentatively referred to the family Taxodiaceae.

Protopinaceans, which represent an interme-

Fig. 3.4. Partial leaf of the bennettite *Otozamites mundae*, a group of plants that is related to the modern palm ferns (Cycadeacea) and went extinct in the Early Cretaceous. Height of the preserved leaf is 3.2 cm.

◁◁

Fig. 3.5. Tips of twigs of *Brachyphyllum*, a common form genus of the Mesophytic. Heigth 2.2 cm.

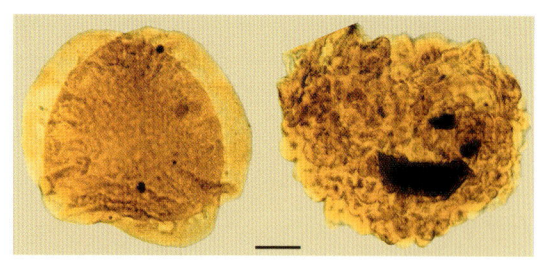

Fig. 3.6. *Callialasporites trilobatus* (BALME 1957) SUKHDEV 1961 (left) and *Cerebropollenites macroverrucosus* (THIERGART 1949) SCHULZ 1967 (right), conifer pollen. *Callialasporites* is regarded as an araucarian, while *Cerebropollenites* probably represents a relative of the taxodiaceans. Scale bar = 10 µm.

Fig. 3.7. Fossil charcoal (fusite) of *Prototaxodioxylon* with asymmetric growth rings and pith tissue (scanning electron-microscopic photograph). Scale bar = 100 µm.

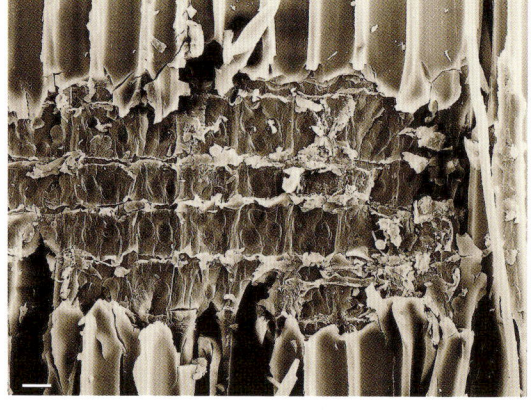

Fig. 3.8. Cross fields of *Prototaxodioxylon* with glyptostroboid pits (scanning electron-microscopic photograph). Scale bar = 10 µm.

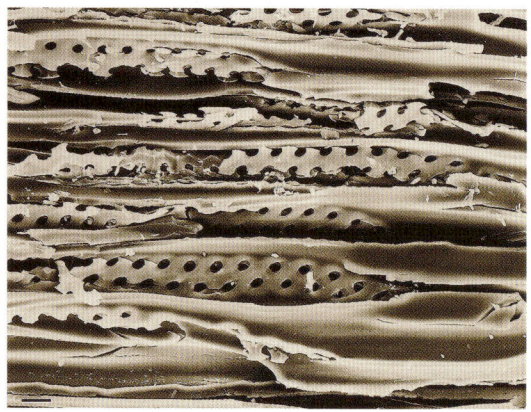

Fig. 3.9. Radial tracheid walls of *Prototaxodioxylon* with typical protopinoid pits (scanning electron-microscopic photograph). Scale bar = 10 µm.

diate form between araucariaceans and pinaceans (Fig. 3.7), are present in the form of fossil charcoal (fusite; identification by H. SÜSS, Berlin). Based on its morphology, it can be referred to the genus *Prototaxodioxylon*, which shows glyptostroboid pits in the cross field (Fig. 3.8) (VOGELLEHNER 1968). Protopinoid pits are found on the radial tracheid walls (Fig. 3.9). Since this combination has not been described so far, the specimen probably represents a new species. Relatively large, bisaccate pollen indicates the occurrence of pinaceans, or at least of close relatives.

Ginkgos

Ginkgos are known from the locality Cabo Mondego with the species *Baiera viannae* (PAIS 1998). However, this group of plants has rarely been documented in the Late Jurassic (PETER 1978). The macro remain mentioned by BRAUCKMANN (1978) cannot be identified precisely, but it might be referable to this taxon (Fig. 3.10).

Reconstruction of the flora

The sparse plant remains only allow limited conclusions about the flora of the coaly swamp that represented the ecosystem of Guimarota. It probably consisted of large woodland swamps with many open ponds and lakes, in which enough plant material was deposited to at least form thin coal seams with a high percentage of minerals. In the lakes, extensive charophyte lawns had formed, and the shores were lined with dense groves of equisetes, which formed a kind of "reed belt" and indicate wet phases in the former peat bog and its surroundings. However, it was not a steady, undisturbed formation of peat; floods due to changes in the courses of rivers and marine ingressions were common (MOHR 1989).

The surrounding forest mainly consisted of conifers, as to be expected in the Middle to Late Jurassic (BEHRENSMEYER et al. 1992), with cheirolepidiaceans and araucariaceans representing the characteristic trees. Pteridosperms and bennettiteans, which had their peak diversity in the Jurassic (MÄGDEFRAU 1968), probably formed the undergrowth. On better drained soils, sandy areas and in the transition to the more elevated hinterland, ferns grew in such abundance that they constitute the second most important element in the microflora of the Guimarota mine. The few relicts of the lycopsid flora – which was diverse in the Carboniferous and played an important role in the formation of coal then – mainly *Lycopodium*, probably grew as herb-like plants in the undergrowth of the forests, or solitary in dry areas, as it is, for example, known from the Early Cretaceous genus *Nathorstiana* from the sandstones of Quedlinburg (MÄGDEFRAU 1968).

On the whole, this paleoflora fits well with other known floras from the Middle to Late Jurassic of western and southwestern Europe, for which the genera *Equisetites*, *Todites*, *Nilssonia* and *Otozamites* were listed as being characteristic by BARALE (1981). PAIS (1974) described a flora of Oxfordian age from Cabo Mondego, in which, similar to the situation found in the Guimarota beds, the species *Equisetites lusitanicum*, *Otozamites mundae* and *Brachyphyllum lusitanicum* together with *Todites falciformis* form a major part. According to PAIS (1974), the flora of Cabo Mondego shows characters that indicate well-developed xeromorph (dry) adaptations, such as, for example, small fern leaflets with thick walls, thick walled bennettite leafs with protected stomata and thick conifer cuticles with sunken stomata. However, such characters should not be regarded as climatic adaptations at face value, since they primarily generally indicate histological or physiological problems in terms of water supply that the plants have to cope with. Thus, the wooden stem of the conifer *Brachyphyllum*, for example, seems to have been small (JUNG 1974), which made the water supply more difficult, especially in old and thus large plants, so that evaporation of water had to be limited as much as possible. The same is true for a Kimmeridgian flora from the vicinity of Lerida (Spain), which also included xerophytic looking forms with small, hard leaves (TEIXEIRA 1956). However, PETER (1978), in a review of Late Jurassic floras from all over the world, concluded that the climate in central and western Europe gradually became drier during the Late Jurassic, which is also supported by sedimentological evidence (HALLAM 1984, 1994). Thus, at least periodically dry periods can also be assumed for the central Portuguese regions.

Glossary

Anthophyta – Totality of all seeding plants
Araucariaceae – a family of conifers, which is known since the Triassic
Caytoniales – an order of seed-ferns (pteridosperms), which is isolated from other pteridosperms because of the organization of its fertile organs, which resembles that of flowering plants (angiosperms)
Cheirolepidiaceae – a family of cypress-like plants that belongs to the conifers
Cyatheaceae – a family of ferns
Gleicheniaceae – a family of ferns
Gnetales – an order of gymnosperms (plants whose seeds are not enclosed in an ovary), which is isolated from other gymnosperms because of the organization of its fertile organs, which resembles that of flowering plants (angiosperms)
Laesura – Y-shaped marker, at which the spore opens when it sprouts
Matoniaceae – a family of ferns
Mesophytic – "Medial time" of the development of plants (Late-Permian to Early Cretaceous)
Osmundaceae – a family of ferns
Perispor – often ornamented outer wall of a spore
Pinaceae – pines; a group of conifers
Pteridosperms – seed ferns; plants with a fern-like morphology, but with true seeds
Schizaeaceae – a family of ferns
Taxodiaceae – a family of conifers ("swamp cypresses")

Fig. 3.10. Partial leaf of a ?gingko, probably from layer FA 11. Height of the remain 3 cm. The remains on the left are tips of the twigs of *Brachyphyllum*.

References

BARALE, G. (1981): La paléoflore jurassique du Jura Français: Étude systématique, aspects stratigraphiques et paléoécologiques. – Documents des Laboratoires de Géologie Lyon **81**: 1-467.

BEHRENSMEYER, A. K.; DAMUTH, J. D.; DIMICHELE, W. A.; POTTS, R.; SUES, H.-D. & WING, S. L. (1992): Terrestrial Ecosystems through Time. – 568 pp., Chicago (University of Chicago Press).

BRAUCKMANN, C. (1978): Beitrag zur Flora der Grube Guimarota (Ober-Jura; Mittel-Portugal). – Geologica et Palaeontologica **12**: 213-222.

FRIIS, E. M. & PEDERSEN, K. (1996): *Eucommiitheca hirsuta*, a new pollen organ with *Eucommiidites* pollen from the Early Cretaceous of Portugal. – Grana **35**: 104-112.

GREGUSS, P. (1968): Xylotomie of the Living Cycads. – 260 pp., Budapest (Akadémiai Kiado).

HALLAM, A. (1984): Continental humid and arid zones during the Jurassic and Cretaceous. – Palaeogeography, Palaeoclimatology, Palaeoecology **47**: 195-223.

– (1994): Jurassic climates as inferred from the sedimentary and fossil record. – In: ALLEN, J. R. L., HOSKINS, B. J., SELLWOOD, B. W., SPICER, R. A. & VALDES, P. J. [eds.] Palaeoclimates and their Modelling: 88-97 London (The Royal Society).

HARRIS, T. M. (1979): The Yorkshire Jurassic Flora. V. Coniferales. – 122 pp., London (Trustees of the British Museum).

HIRMER, M. & HÖRHAMMER, L. (1936): Morphologie, Systematik und geographische Verbreitung der fossilen und rezenten Matoniaceen. – Palaeontographica B **81**: 1-70.

KRAMER, K. U. (1990): Schizaeaceae. – In: KUBITZKI, K. [ed.] The Families and Genera of Vascular Plants. I. Pteridophytes and Gymnosperms: 258-263, Berlin (Springer).

JUNG, W. (1974): Die Konifere *Brachyphyllum nepos* SAPORTA aus den Solnhofener Plattenkalken (unteres Untertithon), ein Halophyt. – Mitteilungen der Bayerischen Staatssammlung für Paläontologie und Historische Geologie **14**: 49-58.

MÄGDEFRAU, K. (1968): Paläobiologie der Pflanzen. – 549 pp., Stuttgart (Fischer-Verlag).

MOHR, B. A. R. (1989): New palynological information on the age and environment of Late Jurassic and Early Cretaceous vertebrate localities of the Iberian Peninsula (eastern Spain and Portugal). – Berliner geowissenschaftliche Abhandlungen A **106**: 291-301.

MOHR, B. A. R. & SCHMIDT, D. (1988.): The Oxfordian/Kimmeridgian boundary in the region of Porto de Mós (Central Portugal): stratigraphy, facies and palynology. – Neues Jahrbuch für Geologie und Paläontologie, Abhandlungen **176**: 245-267.

PAIS, J. (1974): Upper Jurassic plants from Cabo Mondego (Portugal). – Boletim da Sociedade Geológica de Portugal **19**: 19-45.

– (1998): Jurassic plant macroremains from Portugal. – Mémorias da Acádemia das Ciências de Lisboa **37**: 25-47.

PEDERSEN, K. R., CRANE, P. R. & FRIIS, E. M. (1989): Pollen organs and seeds with *Eucommiidites* pollen. – Grana **28**: 279-294.

PEDERSEN, K. R. & FRIIS, E. M. (1986): *Caytonanthus* pollen from the Lower and Middle Jurassic. – Geoskrifter **24**: 255-267.

PETER, H. (1978): Die Florenentwicklung-pflanzengeographische Differenzierung im Zeitraum des Jura. – Schriftenreihe für Geologische Wissenschaften **13**: 1-84.

REMY, W. (1954): Laubfusit, ein Beitrag zur Frage der Fusitbildung. – Glückauf **90**: 64-67.

SCHULTKA, S. (1991): Beiträge zur oberjurassischen Flora des Wiehengebirges. – Geologie und Paläontologie in Westfalen **19**: 55-93.

TEIXEIRA, C. (1956): La flora fossíl de las calizas litografícas de Santa Maria Meya (Lérida). – Publicaciones extranjeras sobre geología de España **9**: 25-36.

VAN ERVE, A. & MOHR, B.A.R. (1988): Palynological investigations of the Late Jurassic microflora from the vertebrate locality Guimarota coal mine (Leiria, Central Portugal). – Neues Jahrbuch Geologie und Paläontologie, Monatshefte **1988**: 246-262.

VAN KONIJNENBURG-VAN CITTERT, H. (1971): In situ gymnosperm pollen from the Middle Jurassic of Yorkshire. – Acta Botanica Neerlandica **20**: 1-96.

– (1981): Schizaeaceous spores in situ from the Jurassic of Yorkshire, England. – Review of Palaeobotany and Palynology **33**: 169-181.

– (1993): A review of the Matoniaceae based on in situ spores. – Review of Palaeobotany and Palynology **78**: 235-267.

VAN KONIJNENBURG-VAN CITTERT, H. & KURMANN, M. H. (1994): Comparative Ultrastructure of living and fossil matoniaceous spores (Pteridophyta) – In: KURMANN, M. H. & DOYLE, J. A. [eds.] Ultrastructure of fossil spores and pollen: 67-86, Kew (Royal Botanic Gardens).

VAN KONIJNENBURG-VAN CITTERT, H. & VAN DER BURGH, J. (1989): The flora from the Kimmeridgian (Upper Jurassic) of Culgower, Sutherland, Scotland. – Review of Palaeobotany and Palynology **61**:1-51.

VOGELLEHNER, D. (1968): Zur Anatomie und Phylogenie Mesozoischer Gymnospermenhölzer, 7: Prodomus zu einer Monographie der Protopinaceae II. Die protopinoiden Hölzer der Jura. – Palaeontographica B **124**: 125-162.

Ostracodes and charophytes from the Guimarota beds

MICHAEL E. SCHUDACK

Remains of ostracodes (small, clam-like crustaceans) and charophytes (algae) are the most common calcareous microfossils in the Guimarota beds and are even found in great quantities in some layers. The ostracodes and charophytes have been the focus of intensive work since the early 1960ies, since it was hoped to obtain valuable information on the dating (biostratigraphy) and paleoecology of this very important vertebrate locality from them. In the meantime, these hopes have been fulfilled (SCHUDACK 2000*).

Ostracodes

Ostracodes are microscopically small arthropods (joint-legged animals), which protect their soft body with two calcareous valves that are arranged on the left and the right sides of the organism and open ventrally. The length of their shells is seldom more than 2 mm, in both fossil and Recent forms. The majority of the almost exclusively aquatic ostracodes live in the benthos, where the animals crawl over the bottom or along plants with their legs, but many of them can swim as well. As all arthropods (e.g. crustaceans) the ostracodes change their skins several times during their lives; after the sloughing, the primarily soft, chitinous valves rapidly calcify again. Because of this, the valves of different larval stages are often very common in the sediment and sometimes even form distinct bodies of rock. The taxonomy and thus the use of ostracodes for biostratigraphy, however, is almost entirely based on adult valves. The diversity of species and abundance of individuals, in the Late Jurassic, as well as today, together with short stratigraphic ranges, low ecological tolerances, and special shapes of valves that reflect their ecology, have made fossil ostracodes one of the most important index- and ecofossil groups.

The pioneer of ostracode research in the Guimarota mine was FRIEDRICH-FRANZ HELMDACH (Freie Universität Berlin), who, after the completion of his PhD-thesis (1968), published on this topic repeatedly (1971, 1973-74, 1991). The ostracodes

Fig. 4.1. *Cetacella inermis* MARTIN 1958, lateral view of the right valve. Length 0.620 mm.

Fig. 4.2. *Cetacella striata* (HELMDACH 1971), lateral view of the left valve. Length 0.735 mm.

from this locality have also been commented on in other, mainly biogeographic works up to the present day (SCHUDACK 1987, 1989, 1996b, 1999, SCHUDACK et al. 1998).

Ten species, representing seven genera within three superfamilies have been identified from the Guimarota mine so far:

Cypridacea: *Cetacella armata* MARTIN 1958, *Cetacella inermis* MARTIN 1958 (Fig. 4.1), and *Cetacella striata* (HELMDACH 1971) (Fig. 4.2).

Most ostracodes of this superfamily live in non-marine environments, in inland waters with freshwater conditions. However, several representatives can tolerate more or less high levels of salinity (brackish conditions). The species of the genus *Cetacella* that are present in Guimarota represent such salinity tolerant forms, which live in oligo- to pliohaline waters with a salinity level of 0 to c. 10 parts per thousand (SCHUDACK 1993b, WEISS 1995, see also SCHUDACK 2000*: Fig. 2.4). In the central European regions, for example in the northern German Kimmeridgian, *Cetacella* is a typical element of brackish faunal associations, where it is often found together with ostracodes of the genus *Macrodentina* and charophytes of the genus *Porochara* (see below) in high abundance. However, these approximately contemporaneous associations usually show a slightly higher salinity than must be assumed for the Guimarota beds.

All of the three species of the genus *Cetacella* that are known from the European Late Jurassic are found in the Guimarota mine. However, their stratigraphic ranges (Late Oxfordian to Late Kimmeridgian for *C. inermis* and *C. striata*, Late Oxfordian to Late Tithonian for *C. armata*) do not allow a more precise dating of the Guimarota beds, other than that they must be of Late Oxfordian to Late Kimmeridgian age (see SCHUDACK 2000*).

Cytheracea: *Theriosynoecum wyomingense* (BRANSON 1935) (Fig. 4.3), *Bisulcocypris pahasapensis* (ROTH 1933), and probably further species of this genus, *Timiriasevia guimarotensis* SCHUDACK 1998 (Fig. 4.4), *Poisia bicostata* HELMDACH 1971 (Fig. 4.5), *Poisia clivosa* HELMDACH 1971, and *Dicrorygma (Orthorygma) reticulata* CHRISTENSEN 1965 (Fig. 4.6).

In terms of the number of species, the Cytheracea are by far the most diverse group of ostracodes today. The representatives of almost all families are exclusively marine, only few can be found in brackish or fresh waters. The Limnocytheridae (called Timiriaseviinae by some authors) represents the only family that is even specialized for these kinds of environments. Limnocytheridae are thus very common in the paleoecologically very variable, mainly limnic to brackish sediments of Guimarota, for example the genera *Theriosynoecum* (Fig. 4.3), *Bisulcocypris* and *Timiriasevia* (Fig. 4.4) (see also SCHUDACK 2000*: Fig. 2.4). All three genera are also widely distributed in non-marine sediments of Kimmeridgian age in Central Europe and North America, with especially close faunal relationships between North America and the Iberian Peninsula on the species level (Fig. 4.7). The reasons for this are the less favorable reproduction- and migration-strategies of the Cytheracea, if compared with the cypridaceans (SCHUDACK 1996b): the species *Theriosynoecum wyomingense*, *Bisulcocypris pahasapensis* and *Timiriasevia guimarotensis*, which are common in both the Guimarota mine, as well as the Morrison Formation of the U.S. Western Interior, had a non-hermaphroditic reproduction and parental care; furthermore, their eggs were not as resistant against drying out and freezing as those of the cypridaceans (see above), so that these species depended on continents or islands for dispersal. Since the Iberian Peninsula was separated from central and north-western Europe by the Proto North Atlantic Ocean, no faunal exchange of cy-

Fig. 4.3. *Theriosynoecum wyomingense* (BRANSON 1935), lateral view of the right valve. Length 0.970 mm.

Fig. 4.4. *Timiriasevia guimarotensis* SCHUDACK 1998, lateral view of the left valve. Length 0.435 mm.

Fig. 4.5. *Poisia bicostata* HELMDACH 1971, lateral view of the right valve. Length 0.485 mm.

Fig. 4.6. *Dicrorygma (Orthorygma) reticulata* CHRISTENSEN 1965, lateral view of the left valve. Length 0.585 mm.

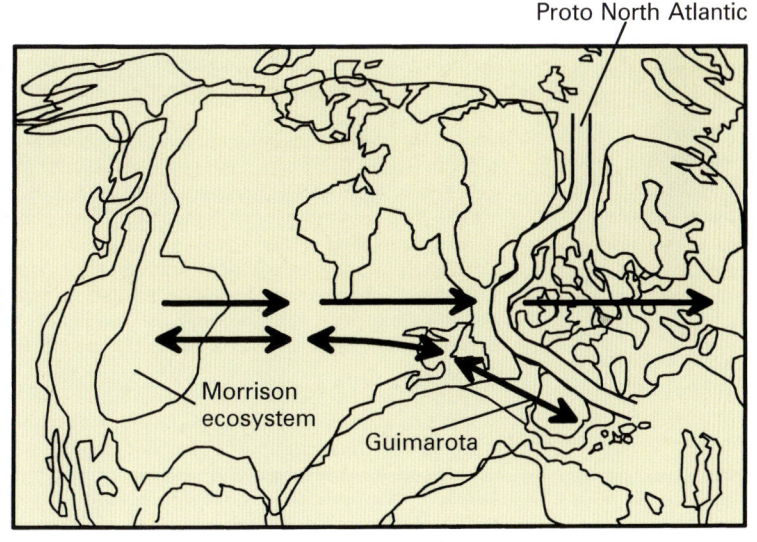

Fig. 4.7. Biogeographic relationships and dispersal patterns of non-marine Kimmeridgian ostracode faunas from the Iberian Peninsula, the Morrison Formation of the western USA, and Central Europe (modified from SCHUDACK 1996b, paleo-coastlines after SMITH et al. 1994).

theraceans was possible between these landmasses. On the other hand, North America was only separated from Iberia by a shallow sea with many islands, which did not represent an obstacle for the dispersion of the Cytheracea (Fig. 4.7). All three species are mainly known from the Kimmeridgian, *B. pahasapensis* extends into the Early Tithonian in the western USA, *T. wyomingense* is already known since the Late Oxfordian.

The genus *Poisia* was first described from the Guimarota mine by HELMDACH (1971), who erected two species *(P. bicostata*, Fig. 4.5, and *P. clivosa)*. Since then, it has been found in other sections of central Portugal by HELMDACH (personal communication), but is unknown in all other regions. Thus, this genus seems to be an endemic (only occurring here) for central Portugal. Not much is known about its ecology, but it certainly occurs in fresh waters. Its stratigraphical range seems to be restricted to the Kimmeridgian (see also SCHUDACK 2000*: Fig. 2.4).

Dicrorygma (Orthorygma) reticulata (Fig. 4.6) is a species that is characteristic for brackish waters (preferred salinity is 0.5 to 5 parts per thousand, according to WEISS (1995), see also SCHUDACK 2000*: Fig. 2.4), but is nevertheless common in all layers in the Guimarota mine. Thus, the layers must have been deposited under very variable conditions with a constant marine influence. HELMDACH (1971) described this form still as *Oertliana kimmeridgensis*; SCHUDACK (1987, 1993a) referred to it under the name that was used for this species at this time, *Dicrorygma (Orthorygma) kimmeridgensis,* and dated the Guimarota Formation as late Lower to early Upper Kimmeridgian on the basis of this taxon. Since then, however, WEISS (1995) convincingly referred this species to *Dicrorygma (Orthorygma) reticulata*, which has a considerable impact on the stratigraphic interpretations. *D. (O.) reticulata* is a widespread European species, which is known from the Kimmeridgian to the Tithonian, in the Baltic sea basin even up to the Berriasian (Early Cretaceous). Thus, no precise biostratigraphic dating of the Guimarota Formation is possible on the basis of these forms.

Darwinulacea: *Darwinula leguminella* (FORBES 1855).

The small, elongate, smooth ostracodes of the genus *Darwinula* are known since the Triassic from nonmarine waters, where they range from fresh waters to hypersaline conditions, and are often found in high abundance (see also SCHUDACK 2000*: Fig. 2.4). They are common, but never very abundant in the Guimarota mine. Species of *Darwinula* are useless for stratigraphy. The morphotype *leguminella* alone has been described from the entire Late Jurassic and even the Early Cretaceous (SCHUDACK 1994).

Charophyta

Charophyte algae are widely distributed in fresh- and brackish waters today, if the water quality is sufficiently good. Due to many investigations, the ecological requirements of Recent taxa are generally well known. This is unfortunately not the case with the fossil representatives, which are mainly represented by their calcified female reproductive organs (gyrogonites, Utriculi).

Fig. 4.8. *Porochara westerbeckensis* (MÄDLER 1952) MÄDLER 1955 from the lower coal seam ("Fundflöz"), gyrogonite in lateral view. Length 0.685 mm.

Fig. 4.9. *Porochara fusca* var. *minor* (MÄDLER 1952) MÄDLER 1955 from the lower coal seam ("Fundflöz"), gyrogonite in lateral view. Length 0.315 mm.

Fig. 4.10. *Porochara fusca* var. *minor* (MÄDLER 1952) MÄDLER 1955 from the lower coal seam ("Fundflöz"), gyrogonite in apical view, showing the rosette-shaped opening that is typical for this genus. Diameter is 0.300 mm.

Fig. 4.11. *Mesochara* sp. from the lower part of the upper coal seam ("Ruafolge"), gyrogonite in lateral view. Length 0.410 mm.

Recent charophytes live submerged (completely covered with water) in waters with low or no salinity. Since, among other factors, their growth depends on the light conditions, they are rarely found in deep water (the maximum is 40 m). They usually grow in shallow waters, which are also favourable for the calcification of parts of the plants. Characean belts are often found parallel to the shores of clean, non-polluted lakes. Here, these algae are especially abundant, both in the number of species, and in the number of individuals. The climatic optimum for charophytes is the warm temperate climate, but they also occur from tropical to arctic conditions. Charophytes are found in all kinds of aquatic environments; apart from lakes, they also grow in ponds, ditches, brackish marginal ocean basins, as well as along the shores of slow running rivers, in water reservoirs, and even in thermal waters. A summary of their possible applications in paleoecological studies can be found in SCHUDACK (1993c), a detailed analysis of the salinity requirements, including those of the species found in Guimarota, in SCHUDACK (1996a). The charophyte gyrogonites figured in Figs. 4.9 to 4.11 are discussed in the chapter on the geology and dating of the locality (SCHUDACK 2000*).

References

BRANSON, C. C. (1935): Freshwater invertebrates from the Morrison (Jurassic?) of Wyoming. – Journal of Paleontology **9**: 514-522.

CHRISTENSEN, O. B. (1965): The ostracode genus *Dicrorygma* POAG 1962 from Upper Jurassic and Lower Cretaceous. – Danmarks Geologiske Undersoegelse, II. Raekke **90**: 3-23.

FORBES, E. (1855) In: LYELL, C. (1855): A manual of elementary geology: or, the ancient changes in the earth and its inhabitants as illustrated by geological monuments, fifth edition: 294-297; London (Murray).

HELMDACH, F.-F. (1968): Oberjurassische Süß- und Brackwasserostracoden der Kohlengrube Guimarota bei Leiria (Mittelportugal). – Unpublished PhD thesis, Freie Universität Berlin.- 92 pp., Berlin.

– (1971): Stratigraphy and ostracode-fauna from the coal mine Guimarota (Upper Jurassic). – Memórias dos Serviços Geológicos de Portugal, N.S. **17**: 43-88.

– (1973-74): A contribution to the stratigraphical subdivision of nonmarine sediments of the Portuguese Upper Jurassic. – Comunicãos dos Serviços Geológicos de Portugal **57**: 4-21.

– (1991): Zur Ornamentierung zweier Arten von *Theriosynoecum* BRANSON 1936 (Ostracoda). – Berliner geowissenschaftliche Abhandlungen A **134**: 119-125.

MARTIN, G. P. R. (1958): *Cetacella*, eine neue Ostracoden-Gattung aus dem Kimmeridge Nordwestdeutschlands. – Paläontologische Zeitschrift **32**: 190-196.

ROTH, R. (1933): Some Morrison Ostracoda. – Journal of Paleontology **7**: 398-405.

SCHUDACK, M. (1987): Charophytenflora und fazielle Entwicklung der Grenzschichten mariner Jura/Wealden in den Nord-westlichen Iberischen Ketten (mit Vergleichen zu Asturien und Kantabrien). – Palaeontographica B **204**: 1-180.

– (1993a): Charophyten aus dem Kimmeridgium der Kohlengrube Guimarota (Portugal). Mit einer eingehenden Diskussion zur Datierung der Fundstelle. – Berliner geowissenschaftliche Abhandlungen E **9**: 211-231.

– (1993b): Die Charophyten aus Oberjura und Unterkreide von Westeuropa. Mit einer phylogenetischen Analyse der Gesamtgruppe. – Berliner geowissenschaftliche Abhandlungen E **8**: 1-209.

– (1993c): Möglichkeiten palökologischer Aussagen mit Hilfe von fossilen Charophyten. – Festschrift Prof. W. KRUTZSCH: 39-60, Berlin (Museum für Naturkunde).

– (1996a): Die Charophyten des Niedersächsischen Beckens (Oberjura – Berriasium): Lokalzonierung, überregionale Korrelation und Palökologie. – Neues Jahrbuch für Geologie und Paläontologie, Abhandlungen **200**: 27-52.

– (1996b): Ostracode and charophyte biogeography in the continental Upper Jurassic of Europe and North America as influenced by plate tectonics and paleoclimate. – Museum of Northern Arizona Bulletin **60**: 333-342.

– (1999): Ostracoda (marine/nonmarine) and paleoclimate history in the late Jurassic of Central Europe and North America. – Marine Micropaleontology **37**: 273-288.

– (2000*): Geological setting and dating of the Guimarota beds. – In: MARTIN, T. & KREBS, B. [eds.] Guimarota – a Jurassic ecosystem: 21-26, München (Verlag Dr. F. Pfeil).

SCHUDACK, M., TURNER, C. E. & PETERSON, F. (1998): Biostratigraphy, paleoecology, and biogeography of charophytes and ostracodes from the Upper Jurassic Morrison Formation, Western Interior, U.S.A. – Modern Geology **22**: 379-414.

SCHUDACK, U. (1989): Zur Systematik der oberjurassischen Ostracodengattung *Cetacella* Martin 1958 (Syn. *Leiria* HELMDACH 1971). – Berliner geowissenschaftliche Abhandlungen A **106**: 459-471.

– (1994): Revision, Dokumentation und Stratigraphie der Ostracoden des nordwestdeutschen Oberjura und Unter-Berriasium. – Berliner geowissenschaftliche Abhandlungen E **11**: 1-193.

SMITH, A. G., SMITH, D. G. & FUNNEL, B. M. (1994): Atlas of Mesozoic and Cenozoic coastlines. – 99 pp., New York (Cambridge University Press).

WEISS, M. (1995): Stratigraphie und Microfauna im Kimmeridge SE-Niedersachsens unter besonderer Berücksichtigung der Ostracoden. – Clausthaler geowissenschaftliche Dissertationen **48**: 1-274.

The mollusk fauna of the Guimarota mine

Martin Aberhan, Frank Riedel & Uwe Gloy

Fig. 5.1. View of the underside of an almost monospecific shell layer of *Isognomon (Isognomon) rugosus* (Münster). Late Jurassic of the Guimarota mine. Length of the slab 50 cm.

Bivalves

The Late Jurassic rocks of the Lusitanian basin of Portugal contain a diverse and well preserved bivalve fauna, which has, however, only partially been described so far (e.g. Choffat 1885-88, 1892-93; Fürsich & Werner 1988, 1989). Bivalves are mainly known from nearshore marine depositional areas, where they often dominate the benthic faunas. Paleoecological and geochemical analyses have shown that most of the bivalves of the Lusitanian basin are euryhaline (tolerating variable salinities) forms and that many of them obviously lived in brackish conditions (e.g. Fürsich & Werner 1986; Yin et al. 1995). The bivalve fauna from Guimarota is strikingly poor in terms of species diversity. On the other hand, single species can be massed in some layers, like, for example, in the almost monotypic shell cover of the slab shown in Fig. 5.1.

Isognomon (**Pterioidea**): Shells of *Isognomon (Isognomon) rugosus* (Münster) (Fig. 5.2) are sometimes very well preserved in the Guimarota mine. This bivalve is known from Portugal from the Late Oxfordian to the Portlandian and occurs in nearshore marine areas in the Lusitanian basin and in the Algarve (Fürsich & Werner 1989). *Isognomon rugosus* lived half buried in the sediment, anchored with a byssus, and fed as a suspension feeder on microorganisms, which were filtered out of the water with the help of its gills. Because of the gregarious lifestyle, these bivalves often form extensive layers of densely packed shells of up to 20 cm in size in the areas were they lived.

The accumulation figured in Fig. 5.1 probably shows such a former community, which was buried in its original habitat. The relatively poor preservation of the single individuals of *I. rugosus* is the result of breakage and distortion due to compaction during the diagenesis, rather than that of fragmentation prior to burial in the sediment. During transport in water currents, the shells of clams are usually abraded and broken, the left and right valves are separated, and the shells are sorted according to their size. Thus, the low degree of fragmentation prior to burial, together with the high percentage of clams in which both valves are preserved in articulation and the broad size range of the shells in Fig. 5.1 indicate that reworking and lateral transport did not play a major role in the accumulation of the shells. In other parts of the Lusitanian basin, *I. rugosus* represents a typical, sometimes the dominant faunal element in brackish environments of lagoons and bays. In contrast to the normal salinity of marine waters of c. 36 parts per thousand, the

Fig. 5.2. Specimen of *Isognomon (Isognomon) rugosus* (MÜNSTER) with both valves preserved. Lateral view of the right valve and medial view of a part of the left valve, showing the ligament shelf. Late Jurassic of the Guimarota mine. Height 9 cm.

Fig. 5.3. Specimen of *"Unio"* cf. *alcobacensis* CHOFFAT with both valves preserved. Lateral view of the right valve. Late Jurassic of the Guimarota mine. Length 4 cm.

salinity in brackish conditions is considerably reduced, with faunal associations of *I. rugosus* being regarded as indicators for brachyhaline (salinity between 30 and 18 parts per thousand) conditions (FÜRSICH & WERNER 1986; YIN et al. 1995).

"Unio": A further bivalve from the Guimarota mine, which is interesting in terms of its ecological implications, is *"Unio"* cf. *alcobacensis* CHOFFAT (Fig. 5.3), a relative of the familiar fresh water clam *Unio* that lives in our native rivers and streams. Although its taxonomic identity is somewhat uncertain, due to the rather poor preservation, *"Unio"* cf. *alcobacensis* can undoubtedly be referred to the superfamily Unionoidea. Thus, in analogy to all modern representatives of this group, it is interpreted as a fresh water faunal element. This means either that there was at least some temporary fresh water influence in the sediments of the Guimarota mine, or that the shells were washed into a nearshore marine depositional environment from a nearby freshwater area.

Snails

The snail fauna from the Guimarota mine is also rich in individuals, but poor in terms of the number of species present. Three taxa could be identified so far, which all indicate a nearshore marine environment. The low number of species can partially be explained by the rather poor preservation of the shells, which makes a precise identification impossible in many cases.

Teinostoma **(Caenogastropoda, Rissooidea):** This genus first appears in the Late Jurassic (WENZ 1938-44); it then experiences a relatively rapid evolutionary radiation and a pantropical distribution of its species. The distribution of the putative Recent representatives of this genus (e.g. KAY 1979) is restricted to the warm and temperate regions of the Pacific ocean. There, they live in marine habitats. The assumption that the species of *Teinostoma* from the Guimarota beds also lived in marine conditions, in analogy with the Recent representatives, is supported by finds of the characteristic boring traces of clionid sponges on the shells of the Late Jurassic snails. These sponges can tolerate temporary brackish conditions, but are bound to marine environments (LAWRENCE 1969).

Cryptaulax **(Caenogastropoda, Cerithioidea):** Procerithiid snails are first known from the Early Jurassic (McDONALD & TRUEMAN 1921). They are

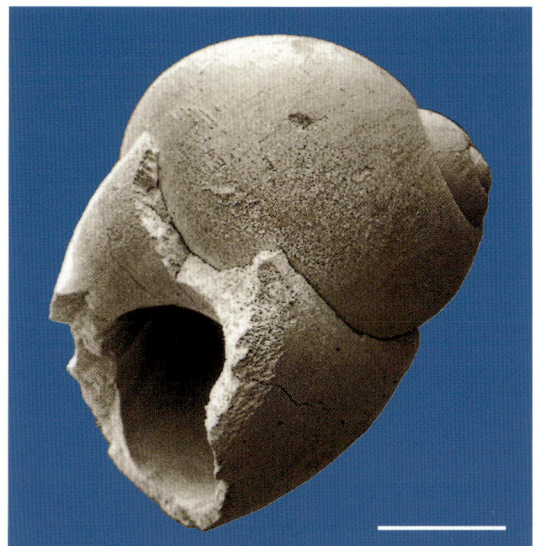

Melampoides (Heterostropha, Ellobioidea): *Melampoides* is a member of the archaeopulmonates, which usually have more or less amphibious habits and often live in environments with marine or brackish influence (e.g. BANDEL & RIEDEL 1998). The Mesozoic taxon *Melampoides* shows close phylogenetic relationships to the ellobiid genus *Melampus*, which lives at least since the Early Cretaceous at the coasts of warm oceans and also invades estuarine areas there (BANDEL & RIEDEL 1998). On both the dry land, as well as under water, the ellobiids prefer soft substrates; the terrestrial habitats must be protected from direct sunlight by overshadowing plants.

The present species from the Guimarota mine was named *Melampoides jurassicus* by BANDEL (1991). In analogy with the low ecological variability of Recent representatives of this group, a amphibious lifestyle at the densely vegetated found with many species in the warm and temperate oceans of the Late Jurassic and are phylogenetically related to the Recent genus *Argyropezza* that lives in indopacific faunas. The latter snail lives on soft substrates in the deeper shelf areas (WILSON 1993). The analogy of the Recent habitat of *Argyropezza* can only partially be applied to the Late Jurassic procerithiid *Cryptaulax* (Fig. 5.6). However, an analysis of Mesozoic faunal associations that contain procerithiids leads to the relatively secure conclusion that *Cryptaulax* lived in marine environments and fed there on algae or detritus.

No larval shells, which might provide evidence on reproductive strategies and dispersal potential, are preserved in the fossil shells.

Fig. 5.4. Adult shell of *Teinostoma* sp. from the Guimarota mine with characteristic columellar shell. Scale bar = 2 mm.

Fig. 5.5. *Teinostoma* sp. from the Guimarota mine: apical view of the early ontogenetic shell. Scale bar = 200 µm.

Fig. 5.6. Adult shell of *Cryptaulax* sp. from the Guimarota mine, showing the characteristic procerithiid shell morphology. Scale bar = 1 mm.

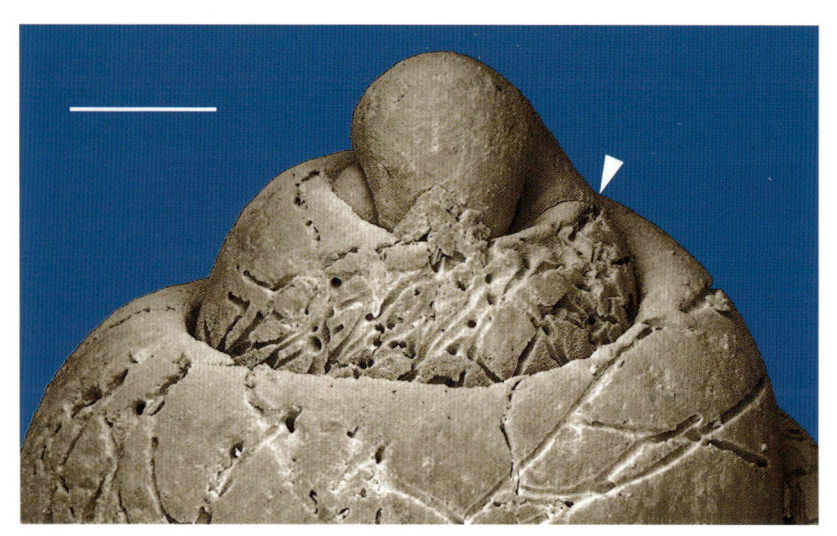

Fig. 5.7. Apex of *Melampoides jurassicus* from the Guimarota mine with anticlockwise spiraling larval shell and clockwise spiraling adult shell, showing boring traces of clionid sponges. The transition from the larval shell to the adult shell is marked by an arrow. Scale bar = 100 μm.

coastline of a warm ocean seems to be likely. Traces of boring activities of clionid sponges and especially the larval shell of *Melampoides jurassicus* (Fig. 5.7) indicate that marine conditions must have been present in the immediate vicinity, since the larval shell is characteristic for planktotrophic larvae, which swim around in the open ocean to find food for their metamorphosis.

References

BANDEL, K. (1991): Gastropods from brackish and fresh water of the Jurassic-Cretaceous transition (a systematic reevaluation). – Berliner geowissenschaftliche Abhandlungen A **134**: 9-55.

BANDEL, K. & RIEDEL, F. (1994): The Late Cretaceous gastropod fauna from Ajka (Bakony Mountains, Hungary): a revision. – Annalen des Naturhistorischen Museums Wien **96A**: 1-65.

– (1998): Ecological zonation of gastropods in the Matutinao River (Cebu, Philippines), with focus on their life cycles. – Annales de Limnologie **34**: 171-191.

CHOFFAT, P. (1885-1888): Description de la faune jurassique du Portugal – Mollusques Lamellibranches, 2e ordre, Asiphonidae. – Mémoire de Direction des Travaux Geologiques du Portugal: 1-76, Lisbon.

– (1892-1893): Description de la faune jurassique du Portugal – Mollusques Lamellibranches, 1er ordre, Siphonidae. – Mémoire de Direction des Travaux Geologiques du Portugal: 1-39, Lisbon.

FÜRSICH, F. T. & WERNER, W. (1986): Benthic associations and their environmental significance in the Lusitanian Basin (Upper Jurassic, Portugal). – Neues Jahrbuch für Geologie und Paläontologie, Abhandlungen **172**: 271-329.

– (1988): The Upper Jurassic Bivalvia of Portugal. Part I. Palaeotaxodonta and Pteriomorphia (Arcoida and Mytiloida). – Comunicações dos Serviços Geológicos de Portugal **73**: 103-144 [für 1987].

– (1989): The Upper Jurassic Bivalvia of Portugal. Part II. Pteriomorphia (Pterioida exclusive Ostreina). – Comunicações dos Serviços Geológicos de Portugal **74**: 105-164 [für 1988].

HUCKRIEDE, R. (1967): Molluskenfaunen mit limnischen und brackischen Elementen aus Jura, Serpulit und Wealden NW-Deutschlands und ihre paläobiogeographische Bedeutung. – Beihefte zum Geologischen Jahrbuch **69**: 1-263.

KAY, E. A. (1979): Hawaiian marine shells. Reef and shore fauna of Hawaii. Section 4: Mollusca. – Bernice P. Bishop Museum Special Publication **64 (4)**: 652 pp., Honolulu, Hawaii (Bishop Museum Press).

LAWRENCE, D. R. (1969): The use of clionid sponges in paleoenvironmental analyses. – Journal of Paleontology **43**: 539-543.

MCDONALD, A. J. & TRUMAN, A. E. (1921): The evolution of certain Liassic gastropods, with special reference to their use in stratigraphy. – Quarterly Journal of the Geological Society London **77**: 297-344.

WENZ, W. (1938-1944): Gastropoda. Teil 1. Allgemeiner Teil und Prosobranchia. – IN: SCHINDEWOLF, O. H. [ed.] Handbuch der Paläozoologie **6 (1)**: 1639 pp., Berlin (Borntraeger).

WILSON, B. R. (1993): Australian marine shells **1**: 408 pp., Kallaro (Odyssey Publishing).

YIN, J., FÜRSICH, F. T. & WERNER, W. (1995): Reconstruction of palaeosalinity using carbon isotopes and benthic associations: a comparison. – Geologische Rundschau **84**: 223-236.

The fish fauna from the Guimarota mine

JÜRGEN KRIWET

The earth's biosphere is divided into terrestrial habitats (land and air) and habitats of aquatic environments (fresh-, brackish- and marine waters). In comparison with the terrestrial habitats, which only constitute approximately half a percent of the biosphere, the aquatic ecosystems present much more ecological niches, which are occupied by different kinds of fishes. In every body of water, ranging from rain water ponds to the deepest depths of the oceans, fishes live and die. Although tetrapods seem to be more variable than fishes, they do in fact not match the differences in size and morphology seen in fishes. The same is true for insects, the most diverse group of organisms at all in terms of the number of species. In the course of their evolution, they evolved from jawless fishes (agnathans) via armored fishes with a rigid bony exoskeleton (Placodermi) to the bony fishes sensu stricto (Actinopterygii). The bony fishes started out with cartilaginous ganoid fishes (Chondrostei) and evolved in several steps towards the modern teleosteans, which are no less complex in their morphology than the terrestrial mammals.

Teleosteans represent the largest group of bony fishes today and live in almost all aquatic environments, with more than 24,000 species (NELSON 1984). Apart from teleosts, bony fishes are represented by a few Recent species of more primitive forms: the bichirs (Polypteriformes), gars (Lepisosteidae) and bowfins (Amiidae). Teleosts began to diversify since at least the Early Jurassic (ARRATIA 1996), and it is thus not surprising that they are widely distributed in the Late Jurassic, in addition to more basal neopterygians, which are often collectively called ganoids ("Holostei"). The holosteans formerly included fishes with rigid scales that are covered with an enamel-like material (Ganoin). However, the name Holostei is not used in systematics anymore, since this group did not originate from a common ancestor (e.g. PATTERSON 1973). The holosteans include several different groups of bony fishes that show a similar evolutionary level, such as the Halecomorphi, Halecostomi and Ginglymodi. The forms which lead towards the modern bony fishes (Teleostei) are also found in this heterogeneous group.

The fish fauna from the Guimarota mine shows the typical components of such faunas from the Late Jurassic of Europe. Within the cartilaginous fishes (Elasmobranchii), the hybodonts are still dominant over the modern sharks (Neoselachii), and bony fishes are exclusively represented by the higher osteichthyans, the neopterygians ("new-fin-fishes"). The latter differ from the "old-fin-fishes" (Palaeopterygii) in the more advanced morphology of the unpaired fins. Apart from the more modern skeletal anatomy of the unpaired fins, the neopterygians also exhibit greater mobility in the jaws, which enabled them to fill a large number of different ecological niches. This is also one of the reasons for the tremendous radiation of the teleosteans during the Tertiary.

Chondrichthyes (cartilaginous fishes)

Sharks are only known from isolated teeth, scales, and head- and fin-spines in the Guimarota coalmine. This is due to the fact that, in contrast to the situation found in bony fishes, the skeleton of sharks consists of cartilage, which is only rarely preserved, and only under special conditions of fossilization. In total, several hundred isolated teeth and scales (placoid scales) and several dozen fin-spines, but only few head-spines have been found in Guimarota. The high number of teeth can be explained by the tooth replacement modus of sharks, which change their teeth throughout their entire life in more or less regular intervals, independent from the degree of wear that the teeth exhibit.

Hybodontiformes

Hybodontiform sharks were the dominant group of cartilaginous fish in the Guimarota ecosystem. Their origins reach back to the Paleozoic. The oldest known hybodont, *Hamiltonichthys*, comes from the Early Carboniferous of Kansas (MAISEY 1989). The hybodonts included the largest and

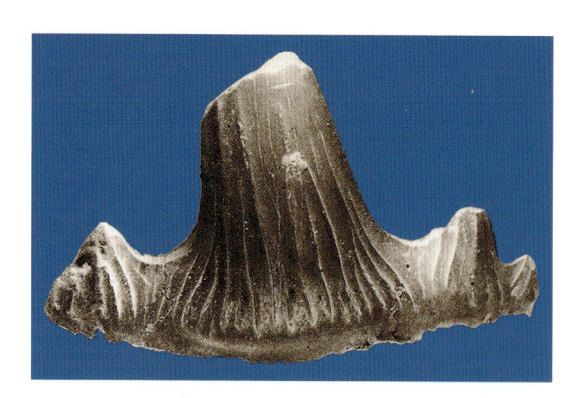

Fig. 6.1. Isolated tooth of *Hybodus* sp. Width 7.3 mm.

Fig. 6.2a. Isolated anterior tooth of *Polyacrodus* sp. Width 2.6 mm.

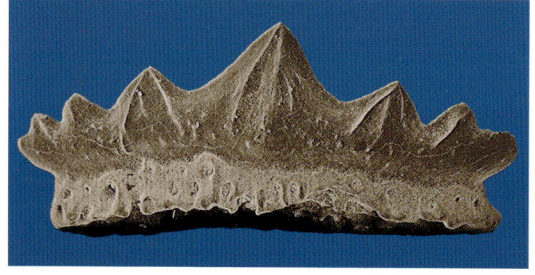

Fig. 6.2b. Isolated tooth of *Polyacrodus* sp. Width 3.2 mm.

than in marine environments. Hybodont sharks differ from modern sharks (Neoselachii) in the more simple histological structure of the teeth and the presence of a primitively built fin-spine in front of each dorsal fin and head spines in male individuals, among other characters. Although they are not directly related to the modern neoselachians, they can be regarded as their sister group (THIES 1983).

Hybodonts are known from the Guimarota mine by teeth of three different genera. Almost all of the teeth found have several distinct cusps (multicuspid teeth). They exhibit a sculpturing, which consists of longitudinal ridges on the labial and lingual sides, which run from the tip of the crown to the basis of the teeth (apico-basally). The larger teeth, which are exclusively represented by isolated, high, and relatively slender tooth crowns, can be referred to the widely distributed genus *Hybodus* (Fig. 6.1), while the smaller and by far most abundant teeth belong to *Polyacrodus* (Fig. 6.2). Several fin-spines of *Polyacrodus* are also preserved (Fig. 6.3). The teeth of *Hybodus* and *Polyacrodus* were arranged in a grasping dentition. Their food probably mainly consisted of invertebrates with a less rigid or no shell (e.g., some clams, slugs), or small fishes. In contrast to the two other hybodonts, the third hybodont, *Asteracanthus*, is very rare in the Guimarota mine; only three remains have been found so far.

The hybodont
Asteracanthus biformatus KRIWET 1995

A. biformatus is one of the largest hybodont sharks from the Guimarota mine. In comparison to the other hybodonts, only little material of this taxon is preserved: a fin-spine, a head-spine and one large tooth (Figs. 6.4 to 6.6). The fin-spine differs from those of other hybodonts in the presence of rounded and star-shaped tubercles, in addition to apico-basally extending ridges on its surface. The teeth formed a crushing dentition, which was used to crack shelled invertebrates. The low-crowned teeth were arranged in several rows in the lower and upper jaws and formed a dense tooth battery. Usually, the teeth are covered by a stout and ornamented enamel-like layer (enameloid), which is similar to the tooth enamel of mammals. However, this layer is missing in the preserved tooth. The channels in the dentine are clearly visible; these channels contained the blood vessels that supplied the tooth with blood. This tooth probably passed the digestive tract of a crocodile.

widest distributed sharks of the Mesozoic. They were primarily marine fishes, but several branches were also adapted to fresh waters. Fresh water hybodonts first occur in the Triassic of Africa, and seem to have had their widest distribution in the Early Cretaceous, with forms such as the genus *Lissodus* (PATTERSON 1966, ANSORGE 1990), which is also widely distributed in the Late Jurassic of Europe, but is absent in Guimarota. In contrast to their marine relatives, fresh water hybodonts were relatively small animals, which rarely exceeded 20 cm in length. They were gradually replaced by the uprising neoselachians, and died out at the end of the Cretaceous. In fresh waters, the hybodonts seem to have resisted the competition pressure of the neoselachians and bony fishes longer

Fig. 6.3. Fin-spine of *Polyacrodus* sp. Height 5.2 cm.

Fig. 6.4. Fin-spine of *Asteracanthus biformatus* KRIWET 1995. Height 9 cm.

Fig. 6.5. Head-spine of *Asteracanthus biformatus* KRIWET 1995. Height 1.6 cm.

Fig. 6.6. Isolated tooth *Asteracanthus biformatus* KRIWET 1995, lateral view. Width 3.5 cm.

In contrast to genera with high tooth crowns, such as *Hybodus* and *Polyacrodus*, hybodonts with low-crowned crushing teeth like *Asteracanthus* lived primarily in marine areas with a normal marine salinity. However, finds from the Purbeck and Wealden of southern England indicate a high salinity tolerance for *Asteracanthus* (e.g. WOODWARD 1895). *Asteracanthus* was probably a rather sluggish animal of nearshore marine areas, which probably also temporarily lived in brackish waters. Due to its considerable size, *A. biformatus* did not have many enemies in the ecosystem of Guimarota. A predator that was potentially able to catch a fish of this size is the marine crocodile *Machimosaurus*, the remains of which have also been found in the coals of Guimarota.

Neoselachians: rare faunal elements

Advanced cartilaginous fishes, which phylogenetically represent the dominant neoselachians today, were still relatively rare in the ecosystem of Guimarota; only a few isolated teeth have been found so far. Recent neoselachians include two morphologically very different groups. These are on the one hand the cartilaginous fishes that are commonly called sharks, and on the other hand the rays (Batomorphi). Rays differ from sharks mainly in a dorso-ventrally flattened body, the pectoral fins that are fused to the head, and the ventrally placed gill slits. Whereas sharks are just known from one tooth from Guimarota, rays seems to have been slightly more common. They are represented by at least two species (Fig. 6.7). The systematic position of the shark to which the sole tooth belongs is still uncertain (Fig. 6.8). Whereas KRIWET (1997) identified it as a posterior tooth of a cat shark (Scyliorhinidae), new finds from the

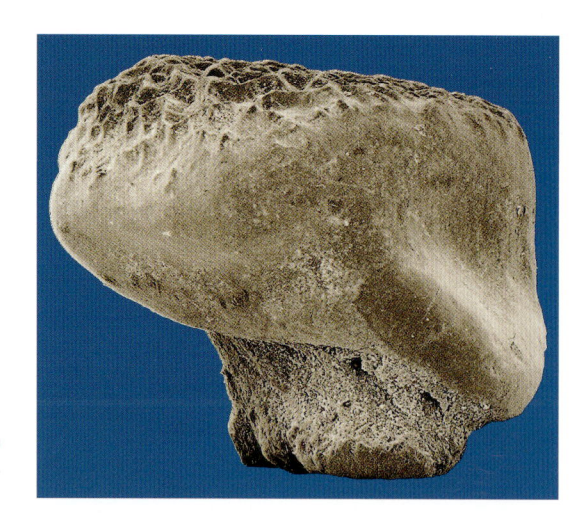

Fig. 6.7. Isolated tooth of an indeterminate ray. Length 1.2 mm.

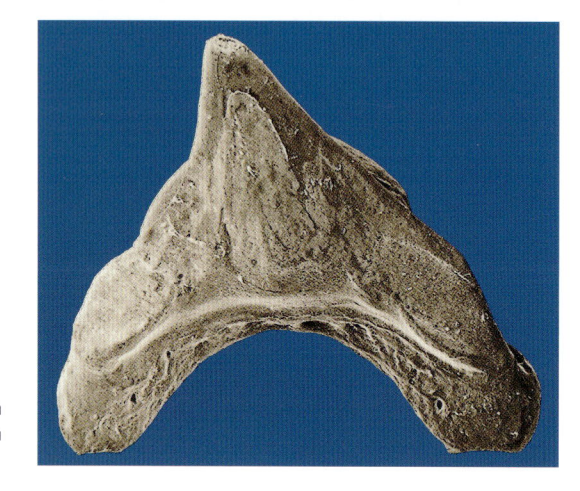

Fig. 6.8. Isolated tooth of a neoselachian. Width 2.4 mm.

Fig. 6.9. Dentalosplenial of an indeterminate macromesiid. Length 2.5 cm.

Late Cretaceous of southern Germany indicate that it might be a carpet shark remain (Orectolobiformes) (F. PFEIL, pers. com.).

Recent carpet sharks live close to the bottom in nearshore shallow waters, and represent one of the phylogenetically oldest group of neoselachians. Morphologically, they are still close to the morphotype found in the original stem-group of neoselachians. However, the reconstruction of the early phylogeny of neoselachians is made more difficult by the recent discovery of isolated teeth from the Late Triassic, which are very similar to teeth of modern neoselachians (CUNY et al. 1998).

Rays probably originated from early, carpet-shark-like cartilaginous fishes. The dorso-ventral flattening of their bodies and the enlargement of the pectoral fins probably represent adaptations to soft substrates (THIES 1983). Another novelty (apomorphy) is their specialized dentition, which is mainly adapted for shelled prey. The teeth are arranged in densely packed rows, one behind the other, and form a crushing or grinding dentition. Rays typically live in shallow nearshore waters, and are often also found in lagoons and the mouths of large rivers. A few species, especially the sting-ray genus *Dasyatis*, are euryhaline and also occur in fresh and brackish waters (COMPAGNO & ROBERTS 1984). However, they return to the oceans for reproduction. One Recent species of rays, though, is adapted completely for a life in the fresh waters of the Nicaragua lake, where it also reproduces.

Osteichthyes (bony fishes)

Although, in contrast to that of the cartilaginous fishes, the skeleton of the bony fishes is well ossified, they are represented mainly by isolated teeth, scales, and indeterminable bones in the Guimarota mine; complete jaws are rare. The reasons for this are the depositional processes in the swamp of Guimarota on the one hand, but also the processing of the sediment during the excavations on the other hand; most of the isolated remains were found in the residue resulting from screenwashing. Only one fish is represented by a partially articulated skull and some scales of the anterior part of the body, which are preserved on a small slab of coal. Isolated teeth of bony fishes can be identified easily, because of their transparent tooth tips, which are formed by acrodine, an enamel-like material. The majority of the fish remains discovered belong to representatives of the "holosteans". These are mainly remains of small to medium sized predatory fishes of shallow coastal waters. Several isolated teeth and jaw remains (Fig. 6.9) can be referred to the macrosemiids, whereas ionoscopids and pachycormids are only represented by isolated teeth. Certain remains of "true" bony fishes (Teleostei) are extremely rare in Guimarota. A few isolated small vertebral centers can be referred to teleosts of uncertain systematic position, based on their degree of ossification and overall morphology.

Shell-crackers with stout armor

The most common fish remains are those of Semionotiformes. However, this order is most probably polyphyletic (it does not originate from a single common ancestor), although WENZ & BRITO (1996) and OLSEN & MCCUNE (1991) listed putative shared derived characters. Since the systematic relationships within this group are still not resolved, though, the name Semionotiformes is here tentatively used in its traditional meaning. Most representatives of this group have rounded teeth, which were arranged more or less densely and irregularly on the mandibles and the palate (Fig. 6.10). Two morphospecies of different size can be distinguished in the extensive material, which includes isolated lateral teeth, lower jaws (dentalosplenials, Fig. 6.11), ganoid scales, as well as isolated skull bones with a typical sculpturing (Fig. 6.12). The rigid ganoid scales were arranged in diagonal longitudinal rows along the body and formed a stout armor for protection against predators (Fig. 6.13). However, this armor greatly restricted the flexibility of the body of these fishes.

The generic determination of the isolated remains of semionotiform fishes from the Guimarota coal mine is problematic, since different genera of semionotiforms had very similar dentitions and scales. The morphology of the teeth, bones and scales is typical for the genus *Lepidotes* AGASSIZ, 1832, ar at least closely related taxa. However, as was the case with the Semionotiformes, the genus *Lepidotes* is not monophyletic and it is in need of revision (THIES 1989). Nineteen species of *Lepidotes* have been described from the Late Jurassic so far, which have a wide geographical distribution from Cuba to Europe. Isolated hookshaped pharyngeal teeth (Fig. 6.14), which were referred to *Lepidotes* by MUDROCH & THIES (1996), and which are relatively common in the Guimarota mine, are also found in derived pycnodontids (see below).

Coral fishes of the Mesozoic

Remains of pycnodont fishes are rare in the Guimarota mine. So far, only few isolated teeth and very fragmentary prearticular dentitions have been found. As was the case for the semionotiforms, two morphologically different groups can be distinguished. Some fragmentary prearticulars with partially preserved dentitions can be referred to a species with the *Coelodus/Proscinetes*-clade, representatives of the most derived group of pycnodonts, and other isolated teeth represent a species of the genus *Macromesodon*. In contrast to the more basal pycnodonts and the semionotiforms, these groups are characterized by a reduction of the scaly armor, so that they had an increased maneuverability.

Both species have transversely broadened grinding teeth, which are arranged in rows on the unpaired vomer and the paired prearticular bones.

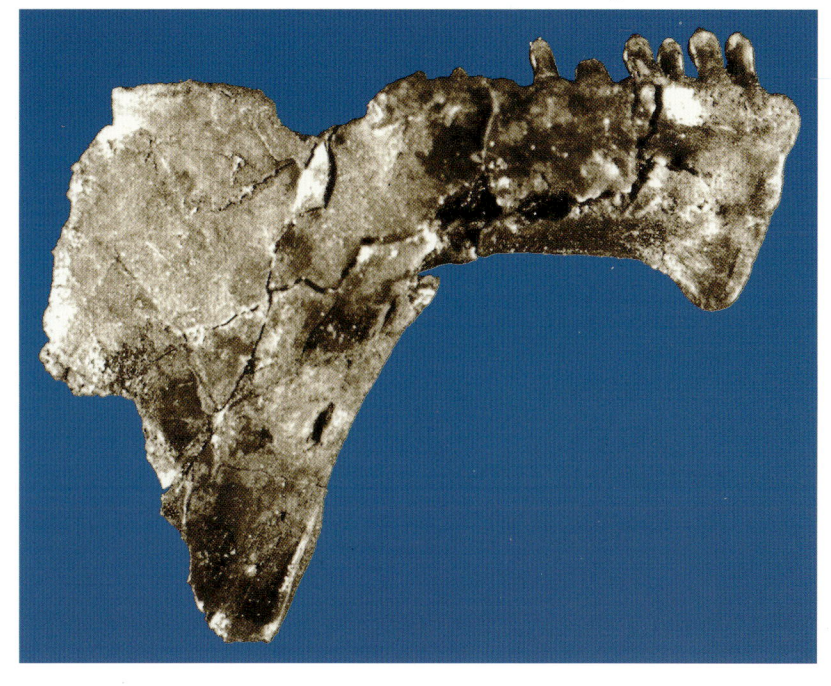

Fig. 6.10. Isolated vomer dentition of *Lepidotes* sp. Length 1.6 cm.

Fig. 6.11. Right dentalosplenial of *Lepidotes* sp. Length 2.8 cm.

Fig. 6.12. Isolated pre-operculum of *Lepidotes* sp. Height 2.7 cm.

Fig. 6.14. Isolated pharyngeal tooth. Height 1.1 mm.

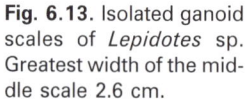

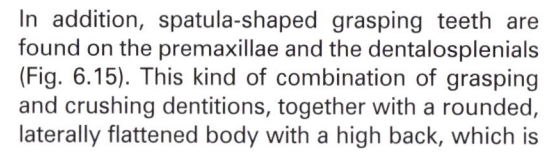

Fig. 6.13. Isolated ganoid scales of *Lepidotes* sp. Greatest width of the middle scale 2.6 cm.

In addition, spatula-shaped grasping teeth are found on the premaxillae and the dentalosplenials (Fig. 6.15). This kind of combination of grasping and crushing dentitions, together with a rounded, laterally flattened body with a high back, which is reminiscent of Recent butterflyfishes, characterize the pycnodonts as highly specialized fishes. The grasping and cutting teeth were not only used to grasp small animals, such as snails and clams, from the ground, but also to graze corals and algae from hard substrates. The shelled or calcified prey was then cracked by the more posteriorly placed grinding dentition. High biting forces could probably be exerted on the prey items. Advanced pycnodonts like *Coelodus* had hook-shaped pharyngeal teeth which were similar to those of *Lepidotes*, and were used to separate the food before swallowing it.

Pycnodonts were typical inhabitants of shallow marine waters, which often lived in areas with algal or coral reefs. With the rise of advanced teleosts with crushing dentitions, like the trigger- and parrotfishes, during the Tertiary, the pycnodonts went extinct.

Fast swimming predators

Representatives of the family Caturidae are known from the Guimarota coal mine from isolated lateral teeth and an incompletely preserved skull. The often recurved teeth have a high tooth neck with a rounded cross-section. The tooth crown is strongly flattened, arrowhead-shaped and shows well-developed cutting edges. Such teeth are characteristic for the genera *Caturus* and *Amblysemius*. In the only preserved skull remain from the Guimarota mine, the palatoquadrate arches and the dentalosplenials, as well as parts of the jaw articulation are preserved in articulation (Fig. 6.16). Behind the skull, some additional body scales of the fish are also preserved on the small slab of coal. These are elongate rounded scales with radial stripes in their anterior parts, which correspond to the type of amioid scales (SCHULTZE 1966, 1996). The typical development of the palatoquadrate arch and the lower jaws, in combination with the amioid scales allows the identification of this fossil as a member of the Caturidae (Amiiformes) (KRIWET in preparation).

The majority of the known caturids were fast swimming predators of the open oceans with a powerful dentition (VIOHL 1987, MUDROCH & THIES 1996). They have an elongate, slender body and range from 10 cm to more than 150 cm in length. Their fossils are known from the Late Triassic to the end of the Early Cretaceous (GARDINER 1993). Phylogenetically, the Caturidae are regarded as the sister-group to the Amiidae; the only Recent representative of the amiids, the bowfin, *Amia calva*, is also the only survivor of the Halecomorphi (sensu PATTERSON & ROSEN 1977). *Amia calva* only occurs in fresh waters in the western USA, especially in the Mississippi area.

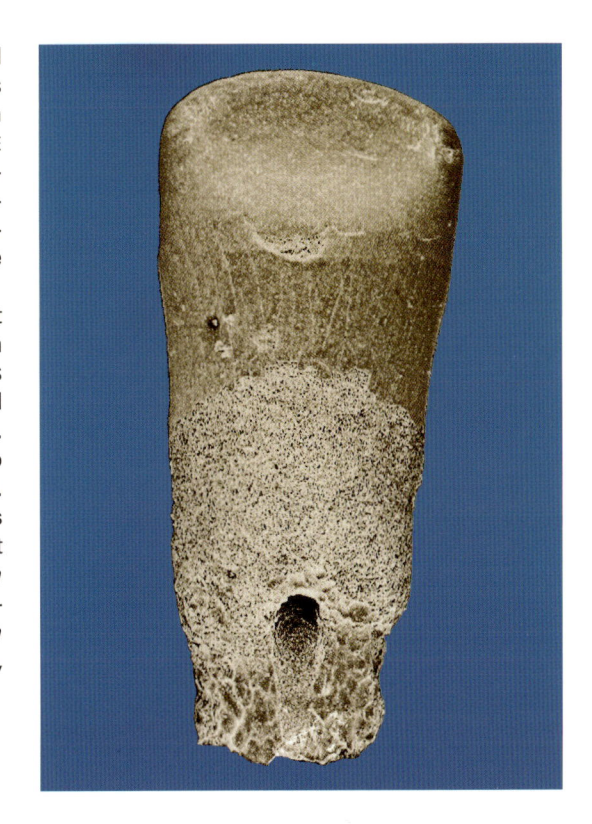

Fig. 6.15. Isolated grasping tooth of *Macromesodon* sp. Height 2.1 mm.

Overview over the fish fauna from the Guimarota mine

Although fishes are only represented by isolated remains, such as bones, scales, dentitions, and teeth in the Guimarota coal mine, a relatively broad faunal spectrum can be recognized. The shark fauna, however, is poorer than that of more or less contemporaneous localities in northern France and Germany in terms of the number of species. The fauna of cartilaginous fishes from Guimarota only consists of three genera of hybodonts and three taxa of neoselachians (KRIWET 1995, 1997, 1998).

In contrast to the cartilaginous fish fauna, the fauna of bony fishes of the Guimarota coalmine is more diverse. However, a determination of the remains on generic level is often impossible, due to the poor preservation of the fossils. The taxa found in the Guimarota mine are listed in the faunal list in the appendix.

Lifestyle of the fishes in the swamp of Guimarota

The coaly sediments of the Guimarota mine are usually regarded as being strongly influenced by terrestrial environments, sometimes even as pure freshwater deposits. This is mainly based on the high percentage of terrestrial tetrapods. An analysis of the sedimentological data and the primarily aquatic organisms has not been carried out so far. It is certain, however, that there was a body of water of unknown dimensions in central Portugal, in the area of Leiria in the Late Jurassic. This body of water was surrounded by a probably dense swamp forest (see MOHR & SCHULTKA 2000*). It is uncertain whether there were any fresh water inlets into the basin.

The hybodonts that lived and searched for food in the ecosystem of Guimarota are usually regarded as euryhaline sharks, which were able to tolerate larger variations in the salinity level of the water. Various fossil rays might have had similar tolerances. Recent rays, especially stingrays and a few blue sharks are able to live in brackish or even fresh waters. However, few Recent ray species that also reproduce in fresh waters are known (e.g. the fresh water ray from the Nicaragua lake and the potamotrygonids of South America). The paleogeographical distribution of *Hybodus* and *Polyacrodus*, e.g. in the Wealden of England and eastern Spain, indicates that these sharks were also able to tolerate varying salinities. The relatively large *Asteracanthus*, however, can be regarded as a typical shallow marine faunal element, based on its known distribution. The same applies to the small shark.

The lack of true predators is noteworthy. All of

Fig. 6.16. Disarticulated skull of a caturid. Maximum width of the coal slab 2.5 cm.

the sharks and rays had durophagous feeding strategies, although the food of *Hybodus* and *Polyacrodus* certainly also included small fishes and invertebrates. The relatively small neoselachian shark probably fed on shelled, as well as unshelled invertebrates and small fishes, or fish larvae, in analogy to its Recent relatives.

In the bony fishes, the enormously high percentage of semionotiform fishes is striking, while the remains of the other fishes present (Macrosemiidae, Caturidae, Pachycormidae and Ionoscopidae) only play a marginal role. At the present state of knowledge, it can be assumed that the majority of semionotids, like the hybodonts, were able to tolerate high variations in salinity, and thus to live in a wide variety of habitats. The genus *Lepidotes* is known from many marginal marine, but also more continentally influenced localities. Because of their rigid and heavy scaly armor, the semionotids were rather sluggish and slow swimming fishes, which had few natural enemies.

Mainly predatory fishes seem to have been present in the former aquatic habitat of the ecosystem of Guimarota. This is especially striking, if the tooth morphology of the separate groups is taken into consideration. The lack of representatives of Recent non-predatory fish groups, however, is not unusual, since the typical non-predatory fishes today, like the carps (Cypriniformes) first appear in the Tertiary. Mesozoic fish faunas, in contrast, are generally dominated by predatory fishes. Within these faunas, the plankton- and plant-eating larvae of the later predatory fishes played the role of the primary consumers in the food chain. The rather small macrosemiids thus represent secondary consumers, while the relatively large caturids and pachycormids are consumers of even higher categories.

Several trophic adaptations can be distinguished within the predatory guild. The semionotids are characterized as durophagous fishes by their crushing dentition, which they used mainly to crush invertebrates with a rigid exoskeleton. A similar lifestyle can be assumed for the pycnodonts with their rows of grinding teeth. If they also grazed on algae and sessile animals, like the

Recent trigger- and parrotfishes, and can thus be regarded as primary consumers, is uncertain. Within the fish fauna, the fast and agile caturids were certainly at the top of the food chain.

Based on the composition of the fish fauna, periodical flooding of the Guimarota basin with marine waters must be assumed. It cannot be determined, whether this happened in the context of the tides, or storms. During such periods, marine fishes invaded the waters of Guimarota, for example the relatively large *Asteracanthus*, possibly in pursuit of pycnodonts. At the same time, the neoselachians also crossed the flooded barriers and got into the ecosystem of Guimarota. Predatory fishes from the open ocean followed these fishes, including the caturids, which, however, did not stay any longer times in these brackish waters. It is difficult to determine, how often and how long these connections to the oceans existed. However, in summary, the picture of a rather heterogeneous fish association can be drawn, which mainly consisted of taxa with a high tolerance for varying salinities and rare marine visitors.

Acknowledgments

The study of the fish material was generously enabled by Prof. Dr. B. Krebs (Berlin), which is gratefully acknowledged. I furthermore thank Dr. T. Martin and Dipl.-Geol. K. Kussius (both Berlin) for critical reading the manuscript. Mrs. W. Harre (Museum für Naturkunde, Berlin) is thanked for the preparation of the color photographs.

References

Ansorge, J. (1990): Fischreste (Selachii, Actinopterygii) aus der Wealdentonscholle von Lobber Ort (Mönchgut/Rügen/DDR). – Paläontologische Zeitschrift **64**: 133-144.

Arratia, G. (1996): The Jurassic and the early history of teleosts. – In: Arratia, G. & Viohl, G. [eds.] Mesozoic fishes – Systematics and Paleoecology: 243-259. – Munich (Verlag Dr. Friedrich Pfeil).

Compagno, L. J. V. & Roberts, T. R. (1984): Marine and freshwater stingrays (Dasyatidae) of West Africa, with description of a new species. – Proceedings of the California Academy of Sciences **43**: 283-300.

Cuny, G., Martin, M. Rauscher, R. & Mazin, J.-M. (1998): A new neoselachian shark from the Upper Triassic of Grozon (Jura, France). – Geological Magazine **135**: 657-668.

Gardiner, B. (1993): Osteichthyes: Basal Actinopterygians. – In: Benton, M. J. [ed.] The fossil record 2: 611-619.- London, Glasgow, New York, Tokyo, Melbourne, Madras (Chapman & Hall).

Kriwet, J. (1995): Beitrag zur Kenntnis der Fisch-Fauna des Ober-Jura (unteres Kimmeridge) der Kohlengrube Guimarota bei Leiria, Mittel-Portugal: 1. *Asteracanthus biformatus* n.sp. (Chondrichthyes: Hybodontoidea). – Berliner geowissenschaftliche Abhandlungen E **16**: 683-691.

– (1997): Beitrag zur Kenntnis der Fisch-Fauna des Ober-Jura (unteres Kimmeridge) der Kohlengrube Guimarota bei Leiria, Mittel-Portugal: 2. Neoselachii (Pisces, Elasmobranchii). – Berliner geowissenschaftliche Abhandlungen E **25**: 293-301.

– (1998): Late Jurassic elasmobranch and actinopterygian fishes from Portugal and Spain. – Cuadernos de Geología Ibérica **24**: 241-260.

Maisey, J. G. (1989): *Hamiltonichthys mapesi*, g. & sp. nov. (Chondrichthyes; Elasmobranchii), from the Upper Pennsylvanian of Kansas. – American Museum Novitates **2931**: 1-42.

Mohr, B. A. R. & Schultka, S. (2000*): The flora of the Guimarota mine. – In: Martin, T. & Krebs, B. [eds.] Guimarota – a Jurassic ecosystem: 27-32, München (Verlag Dr. F. Pfeil).

Mudroch, A. & Thies, D. (1996): Knochenfischzähne (Osteichthyes, Actinopterygii) aus dem Oberjura (Kimmeridgium) des Langenbergs bei Oker (Norddeutschland). – Geologica et Palaeontologica **30**: 239-265.

Nelson, G. J. (1984): Fishes of the world. – 600 pp., New York, Chichester, Brisbane, Toronto, Singapore (John Wiley & Sons).

Olsen, P. E. & McCune, A. R. (1991): Morphology of the *Semionotus elegans* species group from the Early Jurassic part of the Newark Supergroup of eastern North America with comments on the family Semionotidae (Neopterygii). – Journal of Vertebrate Paleontology **11**: 269-292.

Patterson, C. (1966): British Wealden sharks. – Bulletin of the British Museum (Natural History), Geology **11**: 283-350.

– (1973): Interrelationships of holosteans. – In: Greenwood, M. & Patterson, C. [eds.] Interrelationships of fishes: 233-305, London.

Patterson, C. & Rosen, D. E. (1977): Review of ichthyodectiform and other Mesozoic teleost fishes and the theory and practice of classifying fossils. – Bulletin of the American Museum of Natural History **158**: 81-172.

Sauvage, M. E. (1897-98): Vértébrés fossiles du Portugal. Contribution à l'étude des poissons et des reptiles du Jurassique et du Cretacique. – 47 pp., Lisbon (Académie Royal des Sciences).

Schultze, H.-P. (1966): Morphologische und histologische Untersuchungen an Schuppen mesozoischer Actinopterygier (Übergang von Ganoid- zu Rundschuppen). – Neues Jahrbuch für Geologie und Paläontologie, Abhandlungen **126**: 232-314.

– (1996): The scales of Mesozoic actinopterygians. – In: Arratia, G. & Viohl, G. [eds.] Mesozoic fishes – Systematics and Paleoecology: 83-93, Munich (Verlag Dr. Friedrich Pfeil).

Thies, D. (1983): Jurazeitliche Neoselachier aus Deutschland und S-England. – Courier Forschungsinstitut Senckenberg **58**: 1-116.

– (1989): *Lepidotes gloriae*, sp. nov. (Actinopterygii: Semionotiformes) from the Late Jurassic of Cuba. – Journal of Vertebrate Paleontology **9**: 18-40.

VIOHL, G. (1987): Raubfische der Solnhofener Plattenkalke mit erhaltenen Beutefischen. – Archaeopteryx **5**: 33-64.

WENZ, S. & BRITO, P. M. (1996): New data about the lepisosteids and semionotids from the Early Cretaceous of Chapada do Araripe (NE Brazil): Phylogenetic implications. – In: ARRATIA, G. & VIOHL, G. [eds.] Mesozoic fishes – Systematics and Paleoecology: 153-165, München (Verlag Dr. Friedrich Pfeil).

WOODWARD, A. S. (1895): Catalogue of fossil fishes in the British Museum (Natural History), Part III. 544 pp., London (Trustees of the British Museum [Natural History]), London.

The albanerpetontids from the Guimarota mine

MARC FILIP WIECHMANN

So far, mostly albanerpetontids could be identified among the almost 9,000 amphibian bones from the Guimarota mine. Remains of salamanders (cf. *Marmorerpeton*) (S. EVANS, pers. com. 2000) and frogs (discoglossids) are also present, but are not very common. They are subject of the ongoing research (WIECHMANN in prep.).

Occurrence, distribution, and characters of the albanerpetontids

Albanerpetontids are an extinct group of small, terrestrial amphibians with a long tail, short, strong limbs, and an overall appearance that was similar to salamanders (Fig. 7.1). They first appeared in the Middle Jurassic (Bajocian), and their last representatives died out in the Late Tertiary (Miocene). The type locality – the locality that albanerpetontids were first described from – are the Middle Miocene fissure fillings of La Grive-Saint-Alban close to Lyon (ESTES & HOFFSTETTER 1976, FOX & NAYLOR 1982). The distribution of these amphibians reaches from North America (Aptian to Paleocene), via Europe (Bajocian to Miocene) to central Asia (Callovian to Coniacian) (COSTA 1864, SEIFFERT 1969, ESTES & HOFFSTETTER 1976, ESTES 1981, FOX & NAYLOR 1982, NESSOV 1988, MCGOWAN & EVANS 1995, MCGOWAN 1996, MCGOWAN & ENSOM 1997, GARDNER & AVERIANOV 1998). The only find from a southern continent comes from the Early Cretaceous of Morocco (BROSCHINSKI & SIGOGNEAU-RUSSELL 1996). Only isolated bones of albanerpetontids are usually found, like single jaws or cranial elements, vertebrae and limb bones. Only three articulated skeletons have been found so far in the Early Cretaceous lithographic limestones of Las Hoyas/Spain (Barremian) and Pietraroia/Italy (Albian) (COSTA 1864 and MCGOWAN & EVANS 1995).

According to the phylogenetic analyses of MCGOWAN & EVANS (1995) and MCGOWAN (1998), the albanerpetontids represent the sister-group of the Salienta (frogs) and Caudata (salamanders and newts), both of which belong to the Lissamphibia, the "modern" amphibians.

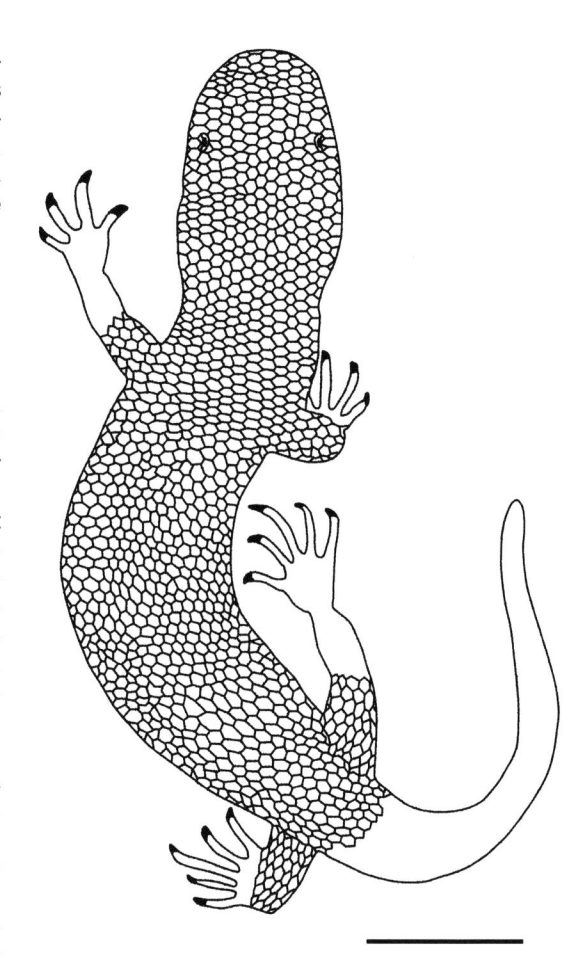

Fig. 7.1. Reconstruction of *Celtedens ibericus* from Las Hoyas (Spain). Scale bar = 10 mm. Redrawn from MCGOWAN & EVANS (1995).

During their long lifetime of some 160 million years, surprisingly little change is found in the albanerpetontids. The characteristic feature of the group is an interdigitating peg-and-socket joint in the symphysis of the mandibles, which makes lower jaws of the albanerpetontids easily identifiable (Fig. 7.2). Further characters that are typical for this group are faintly tricuspate, pleurodont and non-pedicellate teeth (Figs. 7.3 and 7.4), a polygonal ossification pattern on the dermal skull bones, fused frontal bones, and a characteristic atlas-axis-joint in the cervical vertebral column (FOX & NAYLOR 1982; MCGOWAN & EVANS 1995,

Fig. 7.2. Right lower jaw of an albanerpetontid (Gui A 31) from the Guimarota mine with the characteristic peg-and-socket symphyseal joint at the front end. Scale bar = 1 mm.

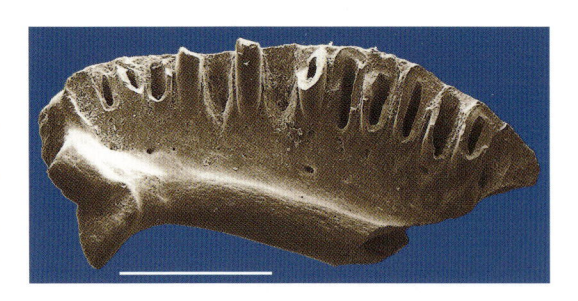

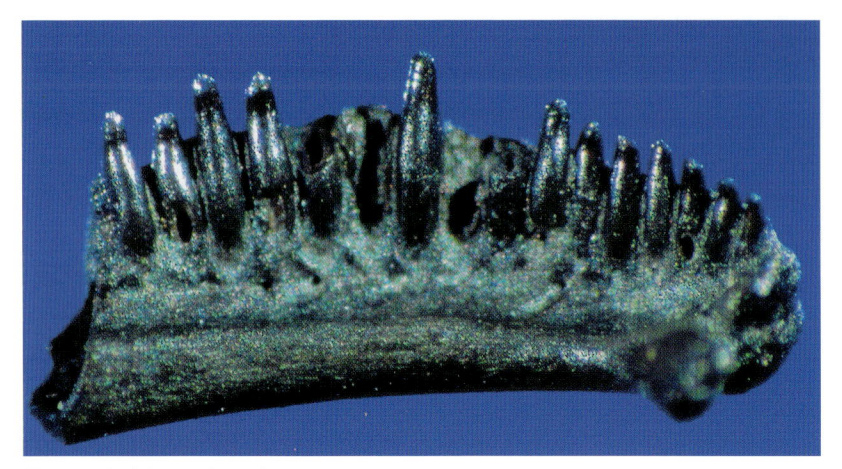

Fig. 7.3. Left lower jaw of an albanerpetontid (Gui A 30) from the Guimarota mine in lingual view with the pleurodont dentition. Length 3 mm.

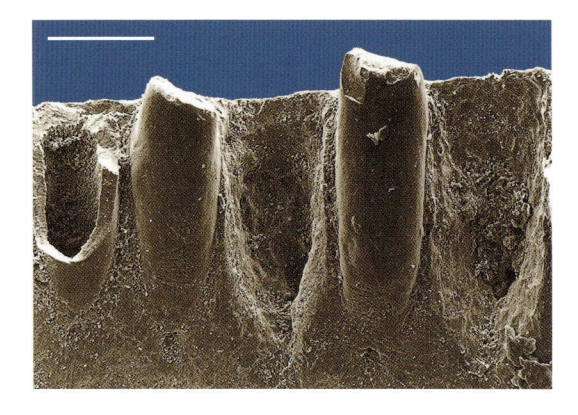

Fig. 7.4. Detail photograph of the pleurodont dentition of the lower jaw figured in Fig. 2. Scale bar = 0.2 mm.

Albanerpetontids from the Guimarota mine

The Guimarota mine has yielded the highest amount of isolated albanerpetontid remains from a single locality. More than 6,500 isolated lower jaws, some 2,300 upper jaw bones (premaxillae and maxillae), more than thirty fused frontals, and many limb-bone fragments, for example the characteristic distal ends of the humeri (Fig. 7.5), were found in the coals of Guimarota. Thus, albanerpetontids are one of the most abundant faunal elements in the ecosystem of Guimarota.

According to MCGOWAN (1998) and my own research, the frontals from the Guimarota mine are very similar to those of the genus *Celtedens* from Las Hoyas/Spain (MCGOWAN & EVANS 1995). A more precise identification of the finds has to await further preparation and detailed investigations of the material (WIECHMANN in prep.).

Lifestyle of the albanerpetontids

The albanerpetontids most probably lived in the ground, as is the case with some salamanders today (e.g. the slender salamander) and especially the gymnophionans (caecilians). Among others, the following characters indicate such a lifestyle: strong joints between the robust cranial bones, fused frontals and partially fused premaxillae, the well ossified and partially fused endocranium, and the atlas-axis joint in the first two vertebrae (FOX & NAYLOR 1982, GARDNER 1999b). The robust skull was probably used as a ram or shovel, to dig a way through the soil. The shape of the occipital condyles and the atlas-axis joint made the necessary movements of the skull and neck possible.

The interdigitigating symphyseal joint, together with the chisel-like, not pedicellate teeth, the robust tooth bearing bones and the tightly joined or even fused premaxillae give hints to the food and feeding strategies. Thus, albanerpetontids probably fed on arthropods with a tough chitinous shell. However, according to FOX & NAYLOR (1982), a durophagous diet (hard-shelled organisms such as clams, for example) can be excluded, because of the chisel-like teeth. The presence of pedicellate teeth in Recent terrestrial salamanders (e.g. tiger salamander) and terrestrial frogs is linked to the use of the tongue in capturing prey. Larger prey which commonly escapes from the sticky tongue is hold by the marginal tooth rows. If the teeth penetrate the prey too deeply or if the prey proves too vigorous, the zone of fibrous tissue allows the tooth crown to break off and the

MCGOWAN & ENSOM 1997). The majority of the Recent, "modern" amphibians – frogs, toads, salamanders, newts, and caecilians – have pedicellate teeth, in which the basis is separated from the crown by an area of fibrous tissue (DUELLMAN & TRUEB 1986).

Two different genera of albanerpetontids *(Albanerpeton* and *Celtedens)* can be distinguished on the basis of the development of the fused frontals (MCGOWAN 1998). Characters of the premaxilla can be used to distinguish the species of the genus *Albanerpeton* (GARDNER 1999a).

prey can be swallowed or released (LARSON & GUTHRIE 1975). Since albanerpetontids do not have pedicellate teeth, it can be assumed that their tongue and marginal tooth rows did not interact in the same way as in their Recent relatives.

ESTES & HOFFSTETTER (1976) argued that the large orbits (eye sockets) in *Albanerpeton* indicate large eyes. However, this assumption is questionable, since Recent salamanders and frogs have small eyes in large orbits. Furthermore, if a burrowing lifestyle is accepted (FOX & NAYLOR 1982), the presence of large eyes in albanerpetontids seems unlikely.

The albanerpetontids probably lived in the humid soil in the vicinity of fresh water ponds, which were used as spawning grounds, in the ecosystem of Guimarota, which was probably similar to the Recent mangrove swamps (e.g. the everglades; mangroves, however, did not exist in the Late Jurassic). MCGOWAN (1996) also compared the habitat represented by the albanerpetontid-bearing beds of Kirtlington/England (Bathonian) with the everglades. Based on the petrography and the micro-invertebrate and microvertebrate fauna, he reconstructed a flat, swampy coastal plain with small streams, lagoons and fresh water lakes. The food of the albanerpetontids probably consisted of small arthropods, which were captured in the burrows, as is the case in Recent terrestrial caudates. The only representative of the terrestrial caudates in the everglades today is the dwarf salamander *(Eurycea quadridigitata)*, which is both diurnal and nocturnal. The habitat of this lungless caudate are swampy areas with low trees, soils covered with conifer needles, protected dark areas under moist tree trunks and the margins of ponds. However, a burrowing lifestyle has not been observed in *Eurycea quadridigitata* (BEHLER & KING 1997).

Taphonomy of the albanerpetontid remains

The question arises, how the large amounts of albanerpetontid bones got into the depositional environment of the coals of Guimarota. It seems possible that terrestrial areas of the swamp were flooded periodically, the digging amphibians drowned and were washed into the areas of deposition with the receding waters. The presence of only isolated elements might have several reasons. After the corpses were washed into the swampy waters, they floated on the water surface for some time, because of the decay gases that formed in the body. In the course of the decay of the corpses, single skeletal elements separated from the body and sank to the bottom of the water. Furthermore, the drifting, bloated amphibians were easy prey for scavengers, such as fishes, crocodiles, and pterosaurs. Between the possible death of an animal in its burrows, or on an "excursion" to the surface, and a flood that washed away the loose upper layers of the soil, including the corpses, there was certainly often enough time for the dead amphibians to be tackled by bacteria or terrestrial scavengers – e.g. small theropods. Thus, the bones were already isolated when they got into the area where they were finally buried.

The dominance of lower jaws and other tooth-bearing elements in the material can be explained by their robust construction, which greatly enhances their chances to be fossilized. In contrast, the relatively delicate postcranial elements are underrepresented, although this is certainly also caused by the fact that they have not been identified yet.

Fig. 7.5. Distal fragment of a right humerus (upper arm bone, Gui A 32). The spherical condyle is in line with the shaft and almost completely round, two typical characters of albanerpetontids. Scale bar = 0.5 mm.

References

BEHLER, J. L. & KING, F. W. (1997): National Audubon Society Field Guide to North American Reptiles & Amphibians. – 743 pp., New York (Alfred A. Knopf).

BROSCHINSKI, A. & SIGOGNEAU-RUSSELL, D. (1996): Remarkable lizard remains from the Lower Cretaceous of Anoual (Morocco). – Annales de Paléontologie (Vert.-Invert.) **82**: 147-175.

COSTA, O. (1864): Paleontologia del Regno di Napoli contenente la desrizione e figure di tutti gli avanzi organici fossili racciusi nel suolo di questo regno. – Atti Accademia Pontaniana, Naples **8**: 1-198.

DUELLMAN, W. E. & TRUEB, L. (1986): Biology of amphibians. – 670 pp., New York, St. Louis, San Francisco (McGraw-Hill Book Company).

ESTES, R. (1981): Gymnophiona, Caudata. – In: WELLNHOFER, P. [ed.] Handbook of Paleoherpetology **2**: 1-115, Stuttgart (Gustav Fischer Verlag).

ESTES, R & HOFFSTETTER, R. (1976): Les urodèles du Miocène de La Grive-Saint-Alban (Isère, France). – Bulletin du Muséum national d'Histoire naturelle **398**: 297-343.

FOX, R. C. & NAYLOR, B. G. (1982): A reconsideration of the relationships of the fossil amphibian *Albanerpeton*. – Canadian Journal of Earth Sciences **19**: 118-128.

GARDNER, J. D. (1999a): The amphibian *Albanerpeton arthridion* and the Aptian-Albian biogeography of albanerpetontids. – Palaeontology **42**: 529-544.

– (1999b): Redescription of the geologically youngest albanerpetontid (?Lissamphibia): *Albanerpeton inexpectatum* ESTES and HOFFSTETTER, 1976, from the Miocene of France. – Annales Paléontologie **85**: 57-84.

GARDNER, J. D. & AVERIANOV, A. O. (1998): Albanerpetontid amphibians from the Upper Cretaceous of Middle Asia. – Acta Palaeontologica Polonica **43**: 453-467.

LARSON, J. H. Jr. & GUTHRIE, D. J. (1975): The feeding mechanism of terrestrial tiger salamanders (*Ambystoma tigrinum melanostictum* BAIRD). – Journal of Morphology **147**: 137-154.

MCGOWAN, G. J. (1996): Albanerpetontid amphibians from the Jurassic (Bathonian) of southern England. – In: MORALES, M. [ed.] The Continental Jurassic. – Museum of Northern Arizona Bulletin **60**: 227-234.

– (1998): Frontals as diagnostic indicators in fossil albanerpetontid amphibians. – Bulletin of the National Science Museum of Tokyo C **24**: 185-194.

MCGOWAN, G. J. & ENSOM, C. P. (1997): Albanerpetontid amphibians from the Lower Cretaceous of the Isle of Purbeck, Dorset. – Dorset Proceedings **120**: 113-117.

MCGOWAN, G. J. & EVANS, S. E. (1995): Albanerpetontid amphibians from the Cretaceous of Spain. – Nature **373**: 143-145.

NESSOV, L. A. (1988): Late Mesozoic amphibians and lizards of Soviet Middle Asia. – Acta Zoologica Cracoviensia **31**: 475-486.

SEIFFERT, J. (1969): Urodelen-Atlas aus dem obersten Bajocien von SE-Aveyron (Südfrankreich). – Paläontologische Zeitschrift **43**: 32-36.

The turtles from the Guimarota mine

THOMAS GASSNER

The oldest fossil turtles lived more than 200 million years ago, and their anatomical construction did not change much since. The characteristic feature of the turtles is their shell, since no other group of animals has this character (MLYNARSKI 1976, GAFFNEY 1990, GAFFNEY & MEYLAN 1988). Even small fragments of turtle shells can be identified as such, since the sutures of the keratinous plates are visible on the bony base. The rigid shell caused several changes in the limbs. The upper arm (humerus) and thigh (femur) are strangely bent in many groups and are thus also diagnostic, although precise identifications on the basis of limb bones are difficult. The same is true for the shoulder blades (scapulae), which exhibit the shape of an L, which is unseen in any other group of vertebrates (MLYNARSKI 1976).

Two different groups can be distinguished within Recent turtles. These are first the cryptodires, which can retract their head in a vertical S-shaped movement of the neck, and second the pleurodires, or side-necked turtles, which bend their neck in a horizontal S-shaped movement. Today, pleurodires occur on former Gondwana continents (southern continents: South America and Australia) (WERMUTH & MERTENS 1996). Apart from the differences in the neck, the two groups can also be distinguished by other characters. In the cryptodires, the pelvis is connected with the theca (the shell of the turtles) by ligaments. In the pleurodires, the pelvis is tightly fused along with the theca along bony sutures. The two groups of turtles also differ in the morphology of the skull. The taxonomic determination of turtles is based on the theca and the morphology of the skull (MLYNARSKI 1976, GAFFNEY & MEYLAN 1988).

The turtles from Guimarota

Turtle remains are relatively abundant in the Guimarota mine. Unfortunately, however, only isolated, more or less fragmentary bony elements of the shell, or remains of the limbs are preserved. The shells of the turtles must have disintegrated prior to burial. This is quite remarkable, since a turtle shell is a very compact structure. Usually, complete thecas are known from other localities, which also protect the bones that are within the shell (BRÄM 1965). The turtle remains from Guimarota probably mainly represent subadult (not fully grown) individuals, the bones of which were still not fully fused, or turtles in which the shell elements did not completely fuse throughout their lives. After the animals died, first the skeleton disintegrated and then the elements of the theca separated from each other. Thus, the single bones of the shell could be transported and were finally buried separately. However, the elements were probably not transported over longer distances, since they do not show any significant abrasion and are generally well preserved. Their internal structure is also virtually unchanged; histological thin-sections of plates of the shell show all the details of the bony structure. The position within the shell can be reconstructed for some plates, which were not damaged during the splitting of the coal. Articulated shell remains, however, are the exception. In this context, a partial shell is noteworthy, which represents approximately a quarter of a complete theca. It is the posterior part of a shell and has originally been described by BRÄM (1973) in a short paper.

Unfortunately, turtles cannot be determined precisely with the help of isolated shell elements, so that the turtles from Guimarota could only be identified on family level. In comparison with other European (Solothurn and Wealden Formation) and North American localities (Morrison Formation), the potential identity of the turtles can be restricted to a few families. According to GAFFNEY & MEYLAN (1988) three families occurred in the Late Jurassic: the Platychelidae (Pleurodira), and the Plesiochelyidae and Pleurosternidae (both Cryptodira).

Three different types of bony shell fragments can be distinguished in the material from the Guimarota mine. Plates with raised ridges and grooves that are parallel to the growth rings of the bone can be referred to the platychelids (BRÄM 1965, MLYNARSKI 1976, GASSNER 1999) (Figs. 8.1 and 8.2).

Fig. 8.1. Fragment of the left posterior part of a carapace of a platychelid turtle (Gui Che 81 [T7]) with three articulated pleurals (coastal plates). Note the suture between the central (plate above the vertebrae) and the lateral (coastal plate). The typical platychelid ornamentation is very well visible in this specimen. Length 6.7 cm.

Another type of bony plates can be referred to the pleurosternids. The shell elements of pleurosternids show an irregular meshwork of small channels, which are flanked by tubercles; towards the periphery, they are arranged in parallel lines that meet the margin of the bone at a right angle (Figs. 8.3 and 8.4). Pleurosternids are a common and widely distributed family in the Late Jurassic (BRÄM 1965, MLYNARSKI 1976, XIANGKUI 1996). This group was identified from the Guimarota mine on the basis of comparisons with complete shells of securely identified pleurosternids from other Late Jurassic localities (OWEN 1853, RÜTIMEYER 1873, OERTEL 1924). Since the neural plates from the Guimarota mine exhibit the typical pattern of tubercles seen in pleurosternids, other fragments with the same pattern can be referred to the same group, based on comparisons with ornamentation patterns in Recent turtles. It should be noted, though, that a similar pattern of tubercles is also found in the genus *Glyptops* from the Morrison Formation of North America. However, the latter genus is closely related to the Pleurosternidae (GAFFNEY 1979). Because of the fragmentary preservation of the turtle material from the Guimarota mine, it cannot be determined whether the remains represent a new, or a known taxon.

The third type of plates exhibits small channels, which are less branched. This pattern almost exclusively occurs on plastron fragments (ventral part of the shell). It is furthermore found on the internal side of peripherals (marginal plates), which externally exhibit the pattern of the platychelids. Thus, it seems very probable that the plastron fragments are referable to the Platychelidae. Such channels also occur on the plastron of Recent turtles, for example in the genus *Batagur*, while the carapace (dorsal part of the shell) exhibits a completely different pattern (GASSNER 1999).

Thin-sections of plates of the shell only show the usual structure of turtle shell elements. The morphology is essentially the same in all three different types of plates. Under a massive bony layer, a more spongy layer is found, which is again followed by a massive layer below. In the shells of pleurosternids, the upper layer is relatively thicker than in platychelids, while the spongy layer seems to be relatively thicker in the platychelids. So far, the latter observations have only be made in the material from the Guimarota mine. Further research is needed to confirm this pattern in turtles from other Late Jurassic localities.

Apart from the remains of subadult turtles, two plates of juveniles are also present. These are a neural (median shell element, above the vertebrae) and a pleural (coastal plate).

Turtle limb bones from the Guimarota mine are unfortunately too poorly preserved for identification (Fig. 8.5). Skull remains have not been found so far.

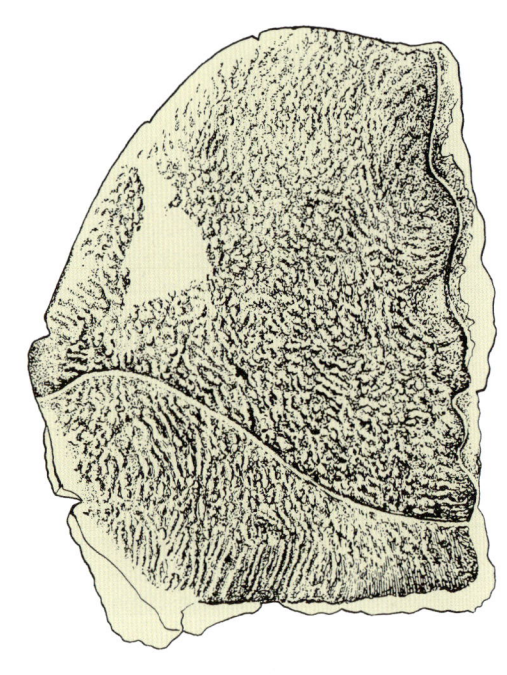

Fig. 8.2. Right epiplastron of a pleurosternid with the typical pattern of tubercles (Gui Che 82 [T8]). The sinuous suture between the keratinous shields, the right gular (first paired keratinous shield of the shell) and the right humeral (second paired shield), is well visible. Length 4.3 cm. From BRÄM (1973).

Fig. 8.4. Typical hexagonal neural (plate above the vertebrae) of a pleurosternid turtle with well preserved pattern of tubercles (Gui Che 18). A groove runs from the posterior part anteriorly; this groove might be a result of an injury. Length 1.7 cm.
◁

◁◁
Fig. 8.3. Right hypoplastron (penultimate plate of the plastron) of a pleurosternid turtle (Gui Che 85 [T9]). The attachment of the contact area between the carapace and the plastron is present in the anterior rim. The suture between the keratinous shields, the femoral (penultimate keratinous shield) and anal (last shield), is very well developed. Length 5.4 cm.

The lifestyle of the turtles from Guimarota

All identifiable turtle remains from Guimarota represent forms that lived in rivers or lakes and had more or less amphibious habits, as judged from a comparison with other localities where such tur-

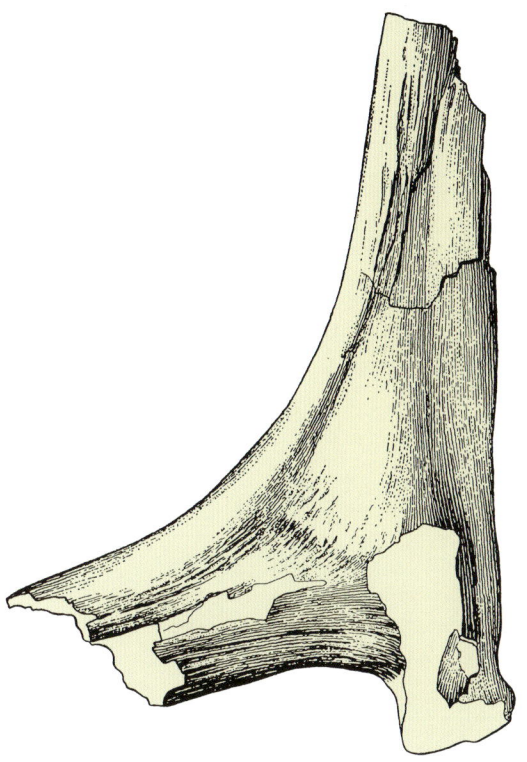

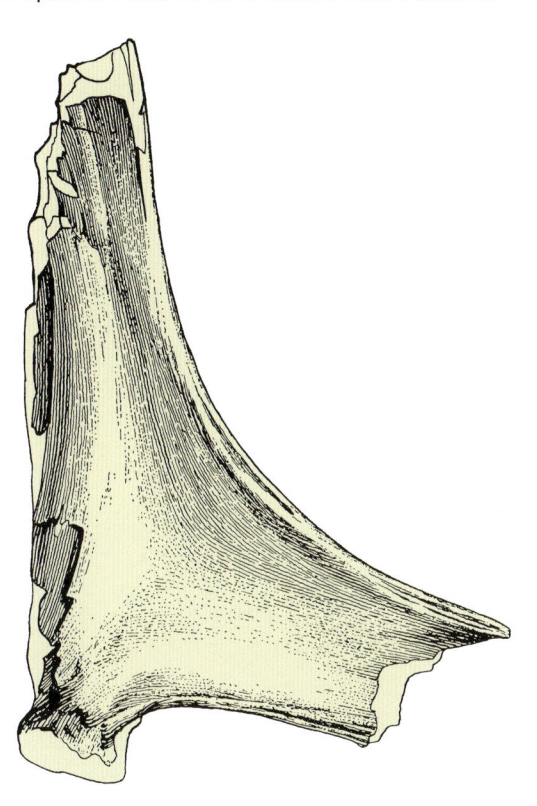

Fig. 8.5. Scapula of an indeterminate turtle from the Guimarota mine (Gui Che 84 [T3]). The scapula of turtles consists of a long scapular blade and a shorter acromion process. Both parts form the L-shape that is typical for turtle scapulae. In the living animal, the scapular blade points dorsally, while the acromion process is directed medially. The glenoid (articular facet) for the humerus (upper arm bone) is placed in the area where the two parts meet. This area is poorly preserved in the element figured. Length 5 cm. From BRÄM (1973).

tles occur (BRÄM 1965). They might also have reached nearshore brackish waters. OERTEL (1924) described the gradual invasion of marine environments by former fresh water turtles in the Late Jurassic of northern Germany. Thus, turtles become more abundant in marine sediments towards the Jurassic-Cretaceous boundary, and their shells become relatively lower. However, the limbs were not adapted for a marine lifestyle yet, since webs between the fingers and toes were probably sufficient for temporary excursions into the open sea. The main habitats of Late Jurassic turtles, though, were rivers, brackish waters and lagoons. True marine turtles first occur in the Cretaceous. Preparations for the invasion of the oceans already happened in the Late Jurassic, when most parts of Europe were covered by shallow seas, from which only isolated islands protruded. The shorelines of this European archipelago were the habitat of the Late Jurassic turtles, which explains the wide distribution of the pleurosternids, for example, in Europe, and, in a broader sense, also in North America, with the closely related glyptonids. Thus, the development in the Late Jurassic represents an important stage in turtle evolution.

References

BRÄM, H. (1965): Die Schildkröten aus dem oberen Jura (Malm) der Gegend von Solothurn. – Schweizerische Paläontologische Abhandlungen **83**: 1-190.

– (1973): Chelonia from the Upper Jurassic of Guimarota mine (Portugal). Contribuição para o Conhecimento da Fauna do Kimeridgiano da Mina de Lignito Guimarota (Leiria, Portugal) III Parte, VII. – Memorias dos Serviços geológicos de Portugal, (nova Sér.) **22**: 135-141.

GAFFNEY, E. S (1979): The Jurassic turtles of North America. – Bulletin of the American Museum of Natural History **162**: 93-135.

– (1990): The comparative osteology of the Triassic turtle *Proganochelys*. – Bulletin of the American Museum of Natural History **194**: 1-263.

GAFFNEY, E. S. & MEYLAN, M. (1988): A phylogeny of turtles. – In: BENTON, M. J. [ed.] The Phylogeny and Classification of the Tetrapods: 157-217, Oxford (Clarendon Press).

GASSNER, T. (1999): Die Schildkröten aus dem Oberen Jura der Kohlengrube Guimarota (Portugal). – Unpublished diploma thesis, Freie Universität Berlin. – 48 pp., Berlin.

MLYNARSKI, M. (1976): Testudines. – In: KUHN, O. [ed.] Handbuch der Paläoherpetologie **7**: 1-130, Stuttgart, New York (Gustav Fischer Verlag).

OERTEL, W. (1924): Die Schildkrötenfauna des nordwestdeutschen oberen Jura. – Paläontologische Zeitschrift **6**: 43-79.

WERMUTH, H. & MERTENS, R. (1996): Schildkröten, Krokodile, Brückenechsen. – 506 pp., Jena (Gustav Fischer Verlag).

XIANGKUI, Y. (1996) The Jurassic Turtles of China. – In: MORALES, M. [ed.] The Continental Jurassic, Museum of Northern Arizona Bulletin **60**: 201-211.

9 The lizards from the Guimarota mine

ANNETTE BROSCHINSKI

Apart from crocodiles, turtles and other reptiles of the Late Jurassic, early representatives of lacertilians, true lizards, are also present in the Guimarota mine. A part of the material that was collected in the end 1950ies and 1960ies has already been published in a monograph by SEIFFERT in 1973. However, more and especially better preserved material was found in the later excavations until 1982, so that a revision of all the lacertilian remains seems necessary (BROSCHINSKI in prep.). The lizard remains from Guimarota represent an important link for our understanding of the older English locality Kirtlington and the slightly younger Purbeck, which has been dated as basal Cretaceous (Berriasian) by ALLEN & WIMBLEDON (1991). Moreover, in comparison with other Late Jurassic and Early Cretaceous localities of the Northern Hemisphere (e.g. Morrison Formation, Solnhofen), they provide further information for the understanding of the evolution of the lacertilians.

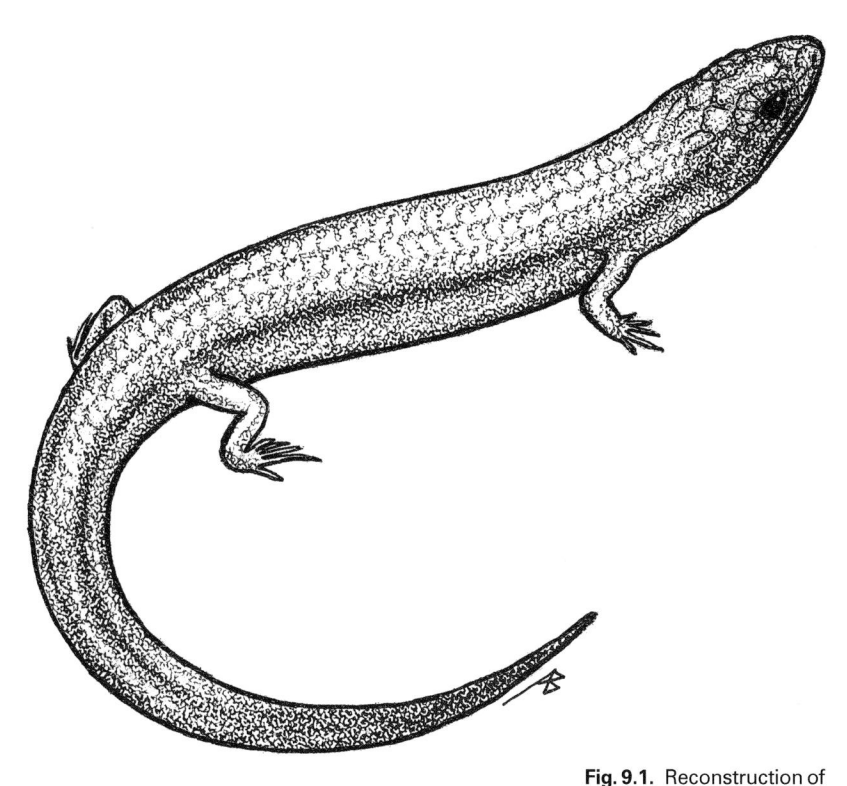

Guimarota lizards

Instead of the expected basal families (Iguania and Gekkota), only the more derived Autarchoglossa, including Scincomorpha and some rare remains of early Anguimorpha, occur in the Guimarota mine. This is similar to the situation found in the Purbeck (HOFFSTETTER 1967; EVANS 1993, 1994) and in the Middle Jurassic locality of Kirtlington (EVANS 1993, 1998b). This is not in accordance with previous assumptions about the evolutionary history of the group and reconstructions of their phylogeny, which thus have rightfully been critically discussed by EVANS (1998b). If the Iguania (iguanas, agamas and chameleons) and Gekkota (geckos and snake lizards) really are the sister groups of the Autarchoglossa (Scincomorpha and Anguimorpha) (ESTES 1991), they would have to be present in fossil localities since the Middle Jurassic. However, the dominance of Scincomorpha is surprising, also in the material from Guimarota. Up to the present day, no gecko remains have been identified from Guimarota with certainty, and representatives of the Iguania are definitively absent. The formerly apparently common geckos from Solnhofen were restricted to a single, questionable species *(Eichstaettisaurus)* by EVANS (1994a), while all other taxa were referred to the Autarchoglossa.

The great abundance of a small, specialized skink-like representative (see below), with possible burrowing (subterraneous) habits, was not to be expected in the Late Jurassic. This taxon might be compared with a highly specialized digging representative of the Scincomorpha from the Early Cretaceous of Anoual/Morocco (BROSCHINSKI & SIGOGNEAU-RUSSELL 1996) and with a small reptile with burrowing habits from the Late Jurassic of Mexico (CLARK & RENE HERNANDEZ 1994). The latter is so strongly specialized that its systematic position cannot be given more precisely than "Diapsida" on the basis of the known skull. The skink-like reptile from Guimarota and Kirtlington (EVANS

Fig. 9.1. Reconstruction of the probably burrowing, skink-like lizard *Saurillodon* from Guimarota.

Fig. 9.2. Reconstruction of the gerrhosaurine-like *Becklesius*.

1998b) is probably related to the Moroccan taxon (EVANS, pers. com.). The ecological specialization of the different groups of lizards obviously began earlier and was more consistent than has been assumed for a long time. The functional obliteration of the few stable characters in the jaws, however, makes a systematic identification of these small reptiles more difficult. So far, they have been referred to the most basal known group of scincomorphs, the Paramacellodidae, based on their occurrence in the same localities and general similarities in the dentitions. In the following, they are tentatively referred to the true skinks (Scincoidea).

The often joint occurrence of two genera of the basal taxon Paramacellodidae, which are gerrhosaurine-like in their appearance, in the Guimarota mine is unusual (Figs. 9.1 and 9.2). Both lack the typical osteodermal ossifications that are regarded as being characteristic for this group – the presence of which, however, has not really been proved for all representatives. Following a proposition by EVANS (1998a), the Paramacellodidae are regarded as the sister group of the Recent African Cordyliformes (gerrhosaurine lizards and girdletailed lizards) here.

Within certain boundaries, the high number of lizard jaws from the Guimarota mine allows an evaluation of the variability within the certainly identifiable remains. Different size stages are mainly found within the genera; these stages have formerly often erroneously been described as distinct species. In a few cases, anomalous development are also found, among others age-dependent proliferous bone growth (synostosis). The same phenomena can be observed in Recent herpetological material, where characters that seem distinctive to the paleontologist can vary between individuals within genera and species (dental foramina: EVANS & BARBADILLO (1997); tooth striations: RICHTER (1994); synostoses: STARCK (1979)). Until recently, new genera and species were described on the basis of such questionable characters.

In the following, the lizards from the Guimarota mine that can be identified with certainty are described ordered according to their abundance, not, as usual, according to their systematic position.

Saurillodon ESTES 1983

The genus *Saurillodon* is a small lizard with a robust skull and a short snout. In the material from Guimarota, many more or less complete jaws are present, but the association of these remains with vertebrae and especially limb bones was so far very difficult. Fortunately, however, one of the rare associated remains from the new excavations could be identified as *Saurillodon*. In this specimen, both skull remains, as well as bones of the postcranial skeleton are found in association on the same slab of coal.

The most complete of the lower jaws of *Saurillodon* from Guimarota shows many characters, but only the most important of these will be briefly described here (Fig. 9.3): the jaw is short and robust; the dentary is very stout and exhibits only 10 to 12 teeth. It represents a short-snouted lizard, which was probably similar to the Recent burrowing forms within the scincomorphs. All Recent burrowing skinks, the skulls of which are often subject to a significant miniaturization sensu RIEPPEL (1989), increase the size of the attachment

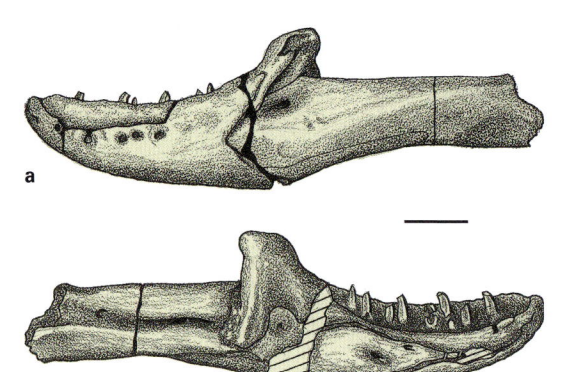

Fig. 9.3. Left mandible of *Saurillodon* (Gui Squ 1); **a** lateral view, **b** medial view. The dentary partially overlaps the coronoid. Scale bar = 1 mm.

Fig. 9.4. Tooth replacement in a right dentary (lingual view) of *Saurillodon* (Gui Squ 3): the 5th tooth is being replaced. Scale bar = 1 mm.

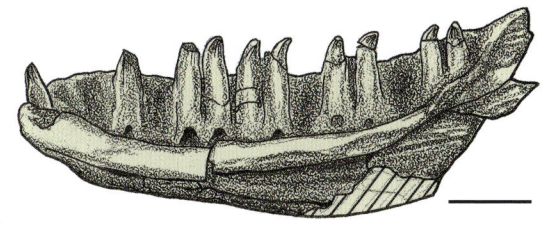

Fig. 9.5. Tooth replacement in a left dentary (lingual view) of *Saurillodon* (Gui Squ 2): well developed resorption pits at the basis of the 4th, 6th, 7th and 9th tooth. Scale bar = 1 mm.

area for the jaw adductor muscles, while at the same time reducing the number of teeth and shortening the dentary. The lower jaw is most similar to the Recent snake skink *Acontias* and the worm-like *Bachia* in terms of its general appearance and morphology. Both lizards are burrowers with reduced limbs. The teeth of these animals only exhibit very fine, weakly developed striations, if striations are present at all. They are replaced from below, as it is typical for scincomorphs, with well-developed, but rather small resorption pits (Fig. 9.4 and 9.5). The upper jaw bones (maxillae) that can be correlated with the lower jaws are also short and have few, but robust teeth (Fig. 9.6).

Both dentary and maxilla exhibit several, sometimes large foramina, which indicates a special sensibilization of the overlying keratinous dermal scales; furthermore, a relatively rough ornamentation is found in the snout. Increased sensibility of the snout is a typical character of strictly ground-dwelling lizards.

Apart from the dentary and the postdentary bone complex, the only associated specimen of *Saurillodon* includes also parts of the skull roof (frontals and an almost complete unpaired parietal). Both of these elements also exhibit a structured surface with many little knobs and bumps, which indicate a well developed ornamentation of the keratinous dermal plates. A well developed foramen is found in the parietal, approximately in the middle of the bone. The preserved boundaries between the keratinous dermal shields, which are discernible on the skull roof elements, indicate that only few, but relatively large keratinous plates were present.

Because of the very obvious functional similarities with Recent burrowing skinks (in the broader sense of the term) in the cranial skeleton, it seemed a safe guess to assume a significant reduction of the limbs (SEIFFERT 1973, EVANS 1996), as well as additional specialisations in the vertebral column of the probably elongate body (EVANS 1998b) in this taxon. The remains of two, apparently only slightly shortened humeri and other limb elements in the associated specimen, however, show that the limbs were only slightly reduced in these lizards. Thus, *Saurillodon* represents a starting point in the evolution of a worm-like body as an adaptation for a subterranean lifestyle.

SEIFFERT (1973) distinguished two species within the genus *Saurillodon*, *Saurillodon proraformis* and *Saurillodon henkeli*. *S. proraformis* is relatively well identifiable – it is represented by the major part of the material, which consists of robust, often coarsely sculptured and massive bones. All of the more gracile material, especially the rather slender and less "boat-like" flexed *Saurillodon*-jaws from the old excavations were referred to the second species, *S. henkeli* by SEIFFERT

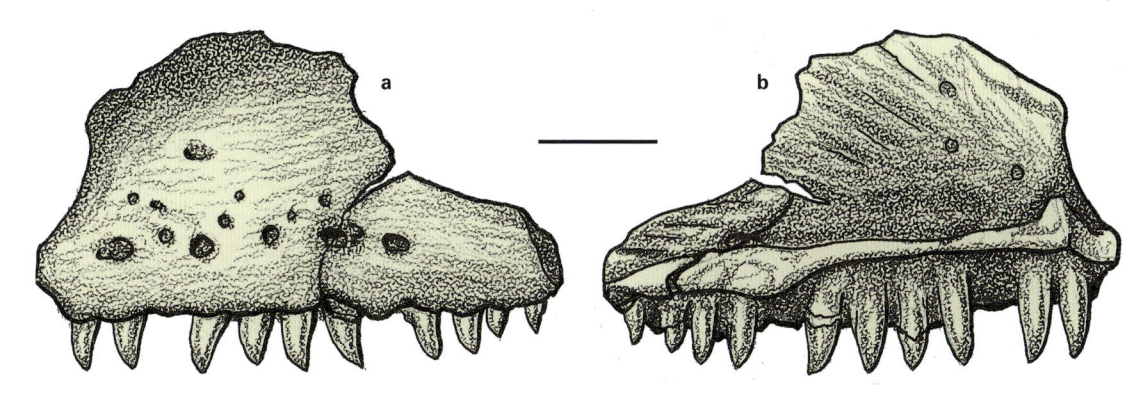

Fig. 9.6. Maxilla of *Saurillodon* (Gui Squ 4); **a** labial view, **b** lingual view. The bone is perforated by many foramina. Scale bar = 1 mm.

Fig. 9.7. Parietal (unpaired) of *Saurillodon* (Gui Squ 5); **a** dorsal view, **b** ventral view. The pineal foramen is placed far posteriorly; the surface is strongly ornamented. Scale bar = 1 mm.

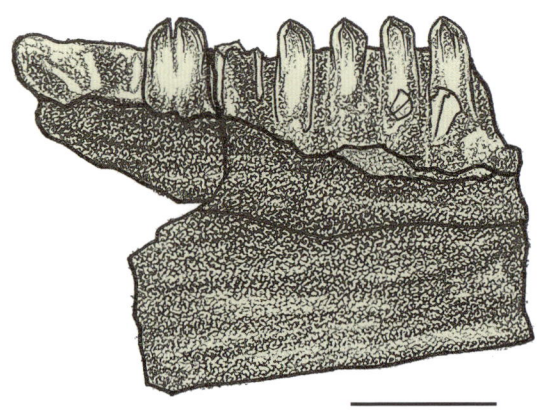

Fig. 9.9. Pathologic developments are sometimes found in the superbly preserved material from Guimarota, for example a dentary fragment of *Becklesius* (Gui Squ 6) with a basally fused "double-tooth", the tips of which point towards each other. Lingual view. Scale bar = 1 mm.

Becklesius ESTES 1983

After *Saurillodon*, the genus *Becklesius* is the second most abundant lizard in the Guimarota mine. *Becklesius* is the most robust representative of the paramacellodids. It is a small lizard, which was probably similar to the gracile Recent gerrhosaurines, with an armor of rectangular keratinous plates – and at least partially with underlying bony plates, the osteoderms.

The Paramacellodidae are a very basal family of the Scincomorpha. They are usually characterized by their osteodermal armor, the morphology of which is mirrored in their Recent sister group, the Cordyliformes (gerrhosaurines and girdle-tailed lizards). The known fossil representatives of the Paramacellodidae are best envisioned as generalized lizards with slightly reduced limbs and a cylindrical body.

Two especially well preserved, large associated specimens of *Becklesius* and other associated specimens without cranial remains that are probably referable to *Becklesius*, are found in the material from Guimarota. The jaws of *Becklesius* bear significantly more teeth than those of *Saurillodon*. Twenty to twenty-five, up to a maximum of 26 teeth are present, which are massive and chisel-like and show a widening of the anterior cutting edge that is typical for this genus. They are closely spaced. The well developed striations on the medial side of the tooth form small inner cusps on the tooth crown, which are especially marked in *Becklesius* (Figs. 9.8 and 9.9).

All cranial bones exhibit a prominent ornamentation, which is, however, less well devel-

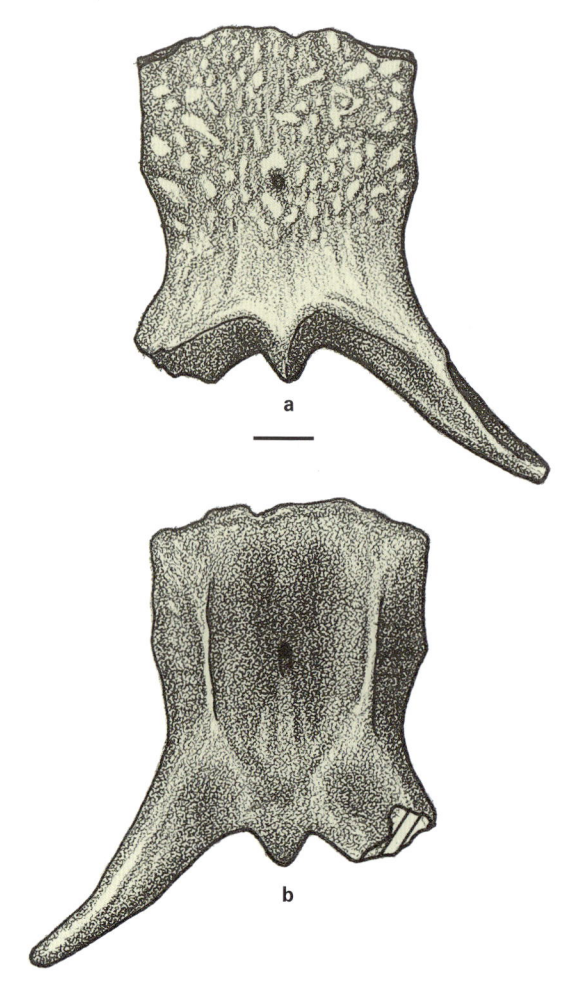

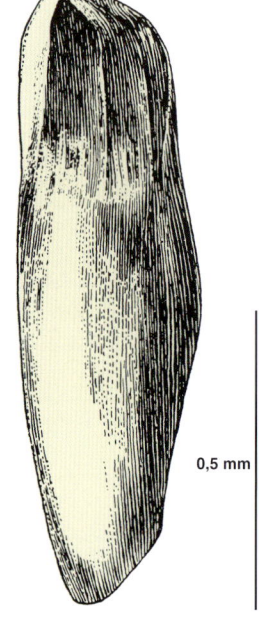

Fig. 9.8. Tooth of *Becklesius*, illustrating the details of the crown morphology (Gui Squ 19 ["Gui A 56"]); lingual view. From SEIFFERT (1973: Fig. 18).

(1973). Both species were also said to differ in the number of teeth, which, however, might rather be due to different age stages. The more numerous and better preserved material from the new excavations will show whether these differences might be age-dependent, or even a sexual dimorphism, rather than taxonomic characters. The systematic position of *Saurillodon* as a representative of the Scincoidea is still debated. Despite many similarities with the Paramacellodidae, *Saurillodon* shows differences to this group in several characters (e.g. the morphology of the parietal, Fig. 9.7). However, the numerous similarities are not surprising in two groups of genera that stand at the basis of the scincomorphs.

oped than in *Saurillodon*. The limbs of *Becklesius* seem to be rather generalized and they are not as reduced as is often postulated for members of the Paramacellodidae.

No traces of osteoderms are found in the preserved associated specimens of *Becklesius*. SEIFFERT (1973) mentioned isolated osteoderms in the material from the old excavations, which, however, cannot be referred to *Becklesius* with certainty: these osteoderms are significantly more robust than those of the specimens of *Paramacellodus* from the Purbeck, and are probably derived from the tail region, since they are especially elongate and have an even and strongly developed keel (RICHTER 1994). Apart from these, some specimens with a pattern of deep groves are striking, which is unusual for this kind of basal scincomorph osteoderms (Fig. 9.10). The articular surfaces in the area where the osteoderms overlap, which have been stated as missing by SEIFFERT (1973), are simply broken off, since the bone is very thin in this area. Because of their well-developed sculpturing and the thickness of the bony plates, it seems probable to refer these osteoderms to *Becklesius*. However, they seem to be rather rare and thus cannot represent remains of extensive osteodermal ossifications of the complete body, since other remains of *Becklesius* are common. *Becklesius* might be compared with the Cordylidae (girdle-tailed lizards) and Gerrhosauridae (gerrhosaurines). Whereas all gerrhosaurids, as well as *Cordylus* have a completely armored body, the South African genus *Pseudocordylus*, for example, only exhibits dorsal armor on the skull and armor in the tail region. However, a completely armored body is the original character state for the whole group. It is questionable, though, if an osteodermal armor really represents a character of all Paramacellodidae. *Becklesius* from the Purbeck, for example, did not have osteoderms, in contrast to *Paramacellodus*, in which, however, the osteoderms are relatively thin, flat and not sculptured. In contrast, *Becklesius* from the Early Cretaceous of Uña had a well-developed and varied osteodermal armor (RICHTER 1994). ESTES (1983) states that *Becklesius* had at least an osteodermal sculpturing of the skull roof, which would again fit well with the armor pattern exhibited by *Pseudocordylus*: flat, thin osteoderms on the skull roof and stout, slender osteoderms with a high keel around the tail might be a feasible model for the reconstruction of *Becklesius* from the Guimarota mine (Fig. 9.11).

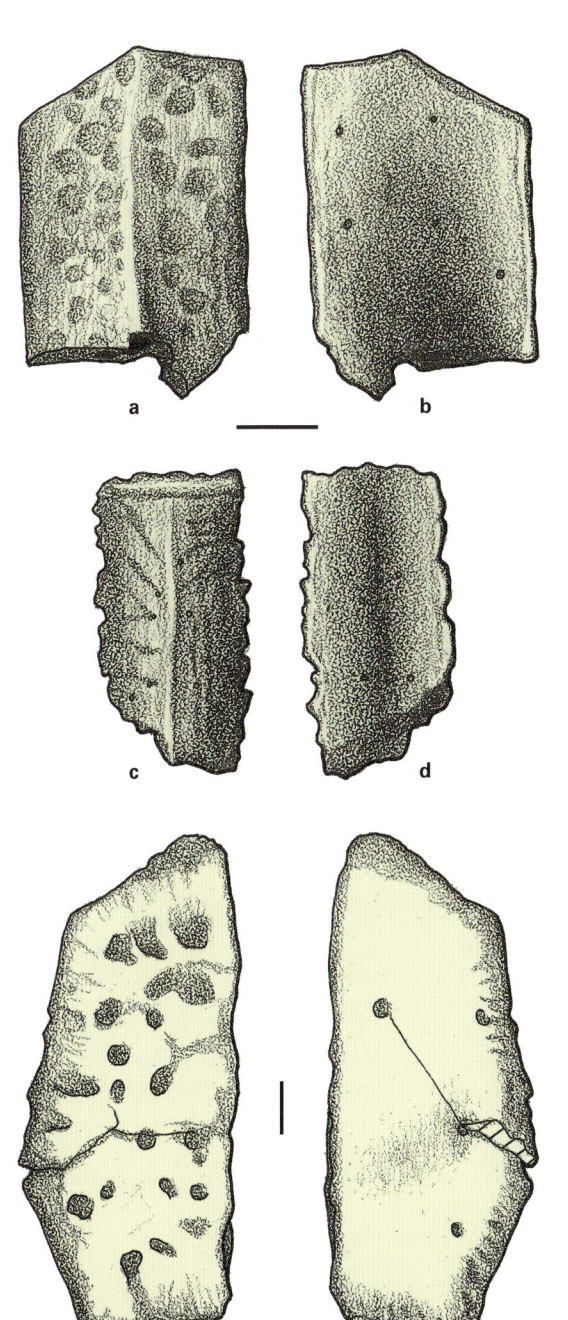

Fig. 9.10. Osteoderms referred to *Becklesius*; **a**, **c** external view, **b**, **d** internal view. a/b probably represents a scute from the thoracic region (Gui Squ 7), c/d represents a caudal scute (Gui Squ 8). Scale bar = 1 mm.

Fig. 9.11. Osteodermal ossification referred to *Becklesius*, probably from the skull roof (Gui Squ 9); **a** external view, **b** internal view. Scale bar = 1 mm.

Paramacellodus HOFFSTETTER 1967

The genus *Paramacellodus*, the name-bearing taxon of the whole family, was so far unknown from the Guimarota mine. Only EVANS (1993) mentioned *Paramacellodus* as "known from Guimarota", however, without further comments. The toothless jaws described by SEIFFERT (1973), most

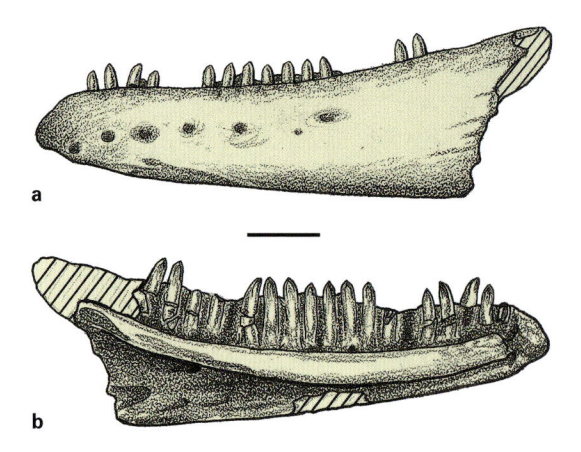

Fig. 9.12. Dentary of *Paramacellodus* (Gui Squ 10); **a** labial view, **b** lingual view. The outer surface is smooth and the teeth are more gracile than those of *Becklesius*. Scale bar = 1 mm.

of which he referred to *Saurillus obtusus*, can now be reevaluated in the light of new, tooth-bearing elements from the new excavations. At least a part of the putative *S. obtusus* remains can probably be referred to the genus *Paramacellodus*, although this needs to be confirmed by comparisons with recent new finds from the Morrison Formation (EVANS 1998a). This is of great importance, since both genera can then be demonstrated to have the same distribution in the Jurassic, as it is the case in the Early Cretaceous localities of Purbeck and Uña. The jaws, which are here referred to *Paramacellodus* for the first time, are much more gracile than those of its sister genus *Becklesius*. Their outer surface is largely smooth and the foramina in the dentaries and maxillae are smaller than in *Becklesius* (Fig. 9.12).

The dentaries have 20 to 24 teeth, the paramacellodid structure of which is evident in the tooth crowns: a well-developed inner cusp is flanked by variously developed striations. The spacing of the teeth is not as tight as in *Becklesius*, but is rather loose. The teeth in the anterior to middle part of the jaws never exhibit a cutting edge that is as well developed as in *Becklesius*. Since the teeth generally appear more slender and sharper than in all other known species of *Paramacellodus* and furthermore the space behind the bony tooth shelf (sulcus dentalis) is unusually wide and deep, the material from Guimarota certainly represents a separate species. Apart from characters in the dentition, the referral of specimens to different genera of paramacellodids is usually based on the presence or absence of ventral muscle scars in the dentary, which is, however, mainly of interest for the specialist. The material from Guimarota shows that this character complex is very variable (Fig. 9.13).

Paramacellodus also exhibits the generalized body shape of the paramacellodids. However, even less can be said about possible osteodermal armor for *Paramacellodus* than for *Becklesius*. Because of the generally smaller and more slender built, thinner and less strongly sculptured osteoderms were to be expected in *Paramacellodus*; however, no such elements have been found so far. No osteoderms are present in the only associated specimen that can be referred to this taxon with certainty. It can be assumed that *Paramacellodus* – as was the case for *Beckelsius* – had a gerrhosaurine-like appearance, possibly also with a *Pseudocordylus*-like armor. Since the known osteoderms from the Purbeck are rather thin and smooth, it is to be expected that such finer or partially reduced osteoderms are less likely to be preserved. As discussed for *Becklesius*, the presence of osteodermal armor is probably less characteristic than it was previously thought. LANG (1991) pointed out that all Recent species of Cordyliformes that exhibit a reduction of the osteodermal armor, are highly specialized crevice-dwellers. Thus, *Pseudocordylus* is better able to squeeze into crevices with its osteoderm-free body, while the head is still protected by the bony armor and can even be wedged between the rocks. It cannot be determined, in how far such adaptations were also of importance for the lizards from Guimarota. *Becklesius* is almost as common as *Saurillodon* in the Guimarota mine, whereas the less common *Paramacellodus* might come from a habitat of the hinterland, not too far away from the Guimarota ecosystem. *Paramacellodus* was probably specialized for less hard food then *Becklesius* with its robust teeth.

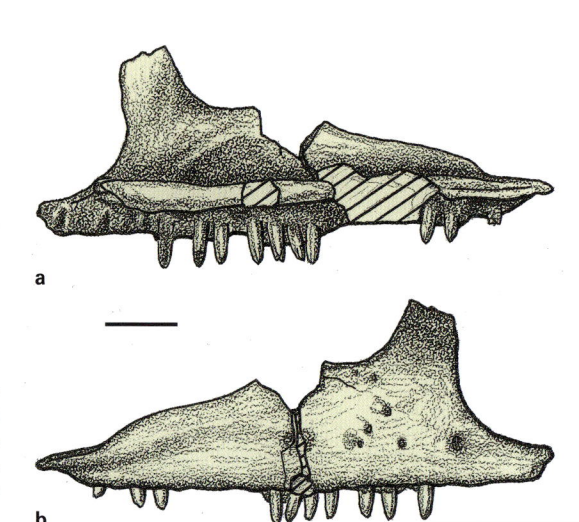

Fig. 9.13. Maxilla of *Paramacellodus* (Gui Squ 11); **a** labial view, **b** lingual view. The nasal process is very high. Scale bar = 1 mm.

Scincomorpha indet.

Another large part of the material referred to *Saurillus obtusus* by SEIFFERT (1973) represents another small scincomorph, which cannot be identified more precisely at present. The material consists almost exclusively of toothless dentaries, which are distinguishable by their better developed and sharply defined facets (for the splenial, angular, etc.) on the inner side, as well as by an extremely wide and deep sulcus dentalis below the tooth row (Fig. 9.14). Some specimens of this undetermined species of scincomorphs exhibit well developed ventral muscle scars in the lower jaws (Fig. 9.15).

Several lower jaws show a tooth replacement pattern with very small, centrally placed resorption pits, in others, the latter are rather large. It is difficult to decide whether the material really represents a single taxon on the basis of the disarticulated specimens. Isolated teeth that were also referred to the "*Saurillus obtusus*-material" by SEIFFERT (1973) may really belong to these jaws, although they are very similar ot the posteriormost five teeth of *Saurillodon*. It is furthermore difficult to recognize remains of the striations in abraded specimens (Fig. 9.16).

Since three new genera and species of small, apparently paramacellodid-like lizards have been described from Kirtlington recently (EVANS 1998b), it can be assumed that at least two of these genera are represented by the present material. Especially the genus *Bellairsia* might be included in the material formerly identified as *Saurillus obtusus*.

Dorsetisaurus HOFFSTETTER 1967

The anguimorph lizard *Dorsetisaurus* has already been described by SEIFFERT (1973) under the generic name *Introrsisaurus* from the Guimarota material. Apart from the true monitor lizards (Varanidae), the lanthanotids, venomous lizards and xenosaurids are included in the Anguimorpha today. ESTES (1983) was able to demonstrate that the remains of *Introrsisaurus* are identical to those of *Dorsetisaurus* from the Purbeck and thus united both genera under the older name *Dorsetisaurus*.

Until the discovery of *Parviraptor* EVANS 1994b, *Dorsetisaurus* was the only anguimorph lizard known from the typical Late Jurassic – Early Cretaceous faunas.

In comparison with the taxa described above, *Dorsetisaurus* is a slightly larger animal. Its dentary bears 18 to 20 teeth, which are widely spaced and slightly recurved. The teeth are considerably flattened transversely and have an anterior and posterior cutting edge (Fig. 9.17). The subdentary shelf is flat and the bases of the teeth exhibit a slight striation, which, however, does not correspond to the folds found in Recent monitor lizards. It is often very weakly developed or even completely absent (Fig. 9.18).

HOFFSTETTER (1967) described parietals from the Purbeck material, which can be referred to

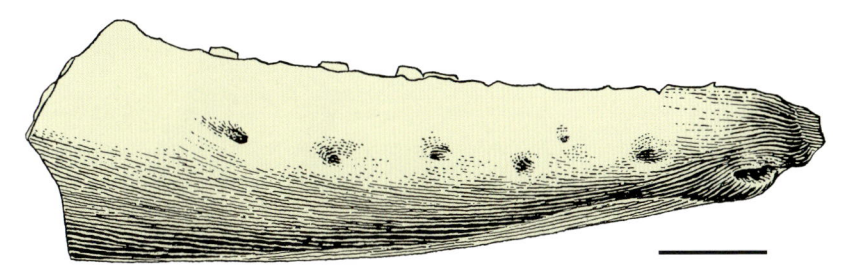

Fig. 9.14. Dentary of a scincomorph (Gui Squ 20 ["Gui 5"]) in labial view. From SEIFFERT (1973: Fig. 22), who referred the specimen to *Saurillus* cf. *obtusus*. Scale bar = 1 mm.

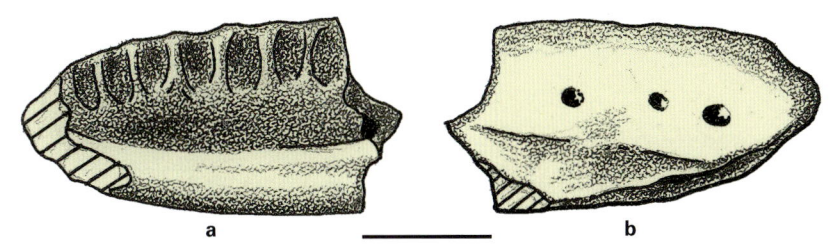

Fig. 9.15. Fragmentary dentary of an undetermined scincomorph (Gui Squ 13); **a** labial view, **b** lingual view. A sharp ridge is developed in the area of the ventral muscle attachments; the sulcus dentalis below the teeth is very broad. Scale bar = 1 mm.

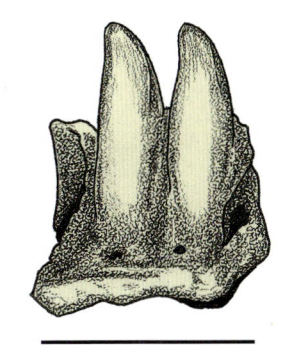

Fig. 9.16. Teeth of an undetermined scincomorph (Gui Squ 14), lingual view. Scale bar = 1 mm.

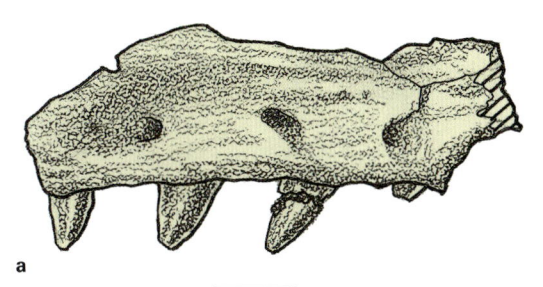

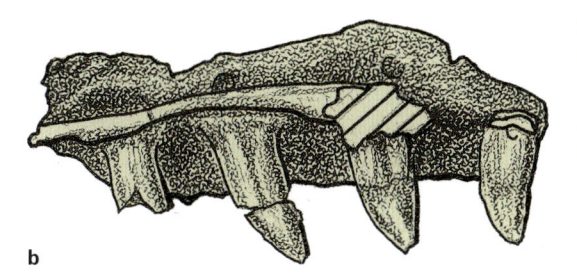

Fig. 9.17. Fragmentary maxilla of *Dorsetisaurus* (Gui Squ 15); **a** labial view, **b** lingual view. The well developed maxillary foramina are placed far ventrally and the widely spaced teeth are transversely flattened. Scale bar = 1 mm.

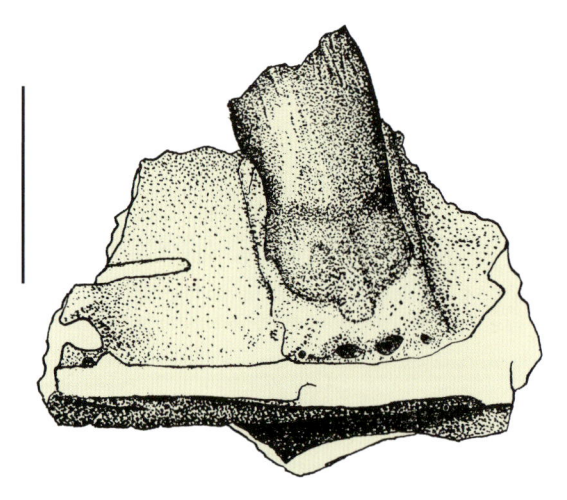

Fig. 9.18. Broken tooth of *Dorsetisaurus* (Gui Squ 21 ["Gui L 232"]) with its pleurodont implantation, which is additionally strengthened, and small resorption pits. From SEIFFERT (1973: Fig. 31). Scale bar = 1 mm.

Fig. 9.20. Anterior half of a dentary of *Parviraptor*, which is preserved in two parts (Gui Squ 18) in lingual view. The teeth are considerably recurved and pointed. Scale bar = 1 mm.

Parviraptor EVANS 1994

The anguimorph genus *Parviraptor*, which was described from the Berriasian (Purbeck) and the Bathonian (Kirtlington) only a few years ago (EVANS 1994b, 1996), has since also been found in other Late Jurassic localities. A few remains of this fascinating, in comparison with the other genera rather large lizard are also present in the material from Guimarota. The present jaw remains clearly exhibit the most important characters (Fig. 9.20 and 9.21).

The widely spaced teeth are significantly recurved, not flattened transversely and end in an extremely slender tip. They have a widened basis, which is indistinctly structured (not folded) and are attached to a narrow subdental shelf. The very large mental foramina are widely spaced. The dentary is straight. The maxilla is also a rather long and slender bone and shows the same tooth implantation as the dentary (Fig. 9.22).

Parviraptor was probably a monitor-like lizard and related to the "monitor-like Anguimorpha" in the broader sense. According to EVANS (1994b), it might be related to the polyphyletic "necrosaurs". Based on its dentition and its relatively large size, if compared to the smaller scincomorphs, *Parviraptor* was probably carnivorous rather than insectivorous and thus also hunted small vertebrates. No material of *Parviraptor* other than the

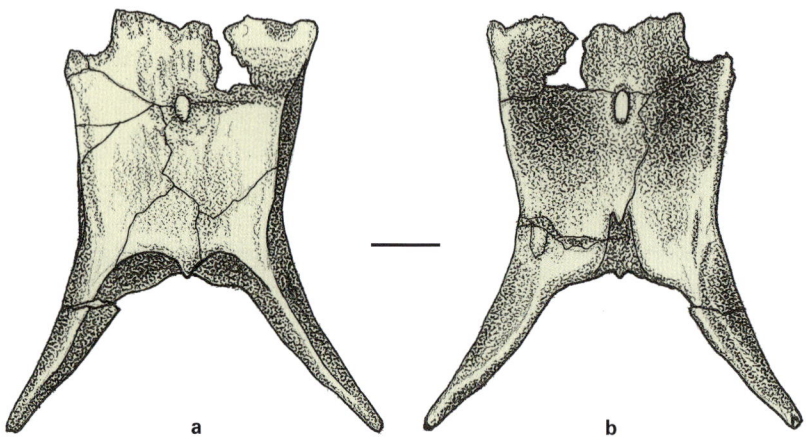

Fig. 9.19. Parietal referred to *Dorsetisaurus* (Gui Squ 16); **a** dorsal view, **b** ventral view. The surface is almost smooth, the two lateral walls are set off from the parietal roof at a distinct angle. Scale bar = 1 mm.

Dorsetisaurus. A few remains of similar parietals are also found in the material from Guimarota; they are smooth and exhibit a large parietal foramen in their middle (Fig. 9.19). In the meantime, *Dorsetisaurus* has also been found in the Late Jurassic dinosaur localities of the Morrison Formation at Como Bluff (ESTES & PROTHERO 1980) and is thus one of the well known and expected faunal elements in Late Jurassic localities of the Northern Hemisphere.

Fig. 9.21. Dentary of *Parviraptor* (Gui Squ 18) in labial view, with pointed, recurved teeth and three large foramina. Length 2.5 cm.

rare jaw elements has been found in the material from Guimarota. EVANS (1994b) illustrated a partial reconstruction of the skull of *Parviraptor estesi*, which gives a good impression of the appearance of this reptile (Fig. 9.23).

The composition of the lizard fauna from the Guimarota mine

The fossil herpetofauna from the Guimarota mine includes five certainly identified genera of lizards within the Squamata. Three of these genera represent Scincomorpha and two belong to the Anguimorpha, whereas remains of geckos are missing. The lizard-taxa discovered in the Guimarota mine so far are listed in the faunal list at the end of this book. The variety in shape that can be demonstrated on the basis of the fossil material is surprisingly broad: from the small, ground-dwelling "skink" with an elongate, worm-like body, via two slightly larger, gracile "gerrhosaurines", which are similar to each other, and a medium-sized predatory lizard, to a very large, long-snouted "monitor", a considerable diversity is present. Thus, the habitat "Guimarota-ecosystem" was obviously generally well suited for small reptiles, with sufficient food sources, ranging from insects for the three scincomorphs to small vertebrates and possibly snails for the two anguimorphs. The ongoing research will certainly provide further important information on the detailed relationships of these lizards from an era in earth history that is crucial for our understanding of the origin of the Recent systematic groups.

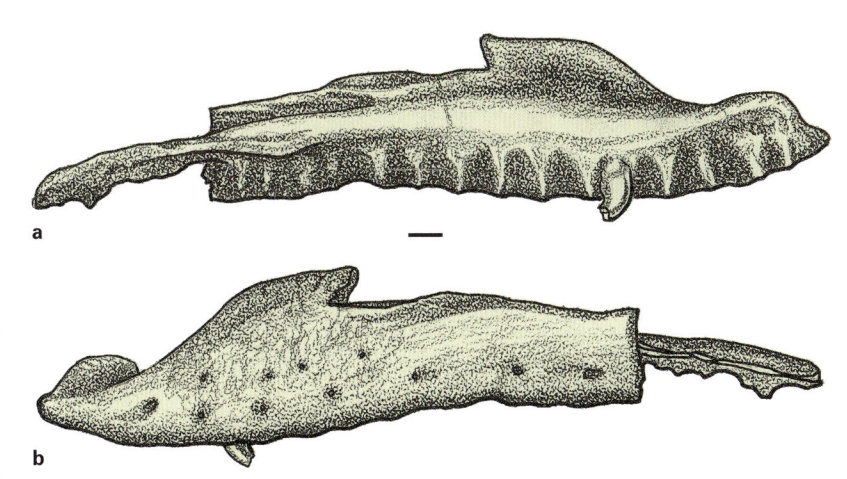

Fig. 9.22. Maxilla referred to *Parviraptor* (Gui Squ 17). The only tooth preserved is recurved up to the break. Scale bar = 1 mm.

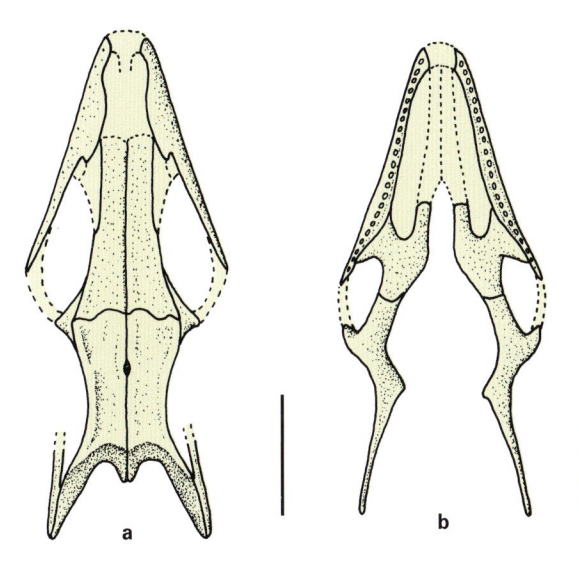

Fig. 9.23. Reconstruction of the skull of *Parviraptor*, dorsal view. From EVANS (1994b). Scale bar = 1 mm.

References

ALLEN, P. & WIMBLEDON, W. A. (1991): Correlation of the Northwest European Purbeck-Wealden (non-marine Cretaceous) as seen from the English type areas. – Cretaceous Research **12**: 511-526.

BROSCHINSKI, A. & SIGOGNEAU-RUSSELL, D. (1996): Remarkable lizard remains from the Lower Cretaceous of Anoual (Morocco). – Annales de Paléontologie (Vert.-Invert.) **82**: 147-175.

CLARK, J. M. & RENE HERNANDEZ, R. (1994): A New Burrowing Diapsid from the Jurassic La Boca Formation of Tamaulipas, Mexico. – Journal of Vertebrate Paleontology **12**: 180-195.

ESTES, R. (1983): Sauria terrestria, Amphisbaenia. – In: WELLNHOFER, P. [ed.] Handbook of Palaeoherpetology **10A**: 1-249; Stuttgart (Gustav Fischer Verlag).

ESTES, R. (1991): Recent perspectives on phylogenetic relationships among reptiles. – Symposium on the evolution of terrestrial vertebrates. – In: GHIARA, G. et al. [eds.] Selected Symposia and Monographs U.Z.I. **4**: 225-254; Modena.

ESTES, R., GAUTHIER, J. & DE QUEIROZ, K. (1988): Phylogenetic relationships within Squamata. – In: ESTES, R. & PREGILL, G. [eds.]: Phylogenetic relationships of the lizard families: 119-281, Stanford (Stanford University Press).

EVANS, S. E. (1993): Jurassic Lizard Assemblages. – Revue de Paléobiologie, Vol. spéc. **7**: 55-65.

– (1994a): The Solnhofen (Jurassic: Tithonian) lizard genus *Bavarisaurus*: New skull material and a reinterpretation. – Neues Jahrbuch für Geologie und Paläontologie, Abhandlungen **192**: 37-52.

– (1994b): A new anguimorph lizard from the Jurassic and Lower Cretaceous of England. – Palaeontology **37**: 33-49.

– (1996): *Parviraptor* (Squamata: Anguimorpha) and other Lizards from the Morrison Formation at Fruita, Colorado. – In: MORALES, M. [ed.] The Continental Jurassic. – Museum of Northern Arizona Bulletin **60**: 243-248.

- (1998a): Paramacellodid lizard skulls from the Jurassic Morrison Formation at Dinosaur National Monument, Utah. – Journal of Vertebrate Paleontology **18**: 99-114.
- (1998b): Crown group lizards (Reptilia, Squamata) from the Middle Jurassic of the British Isles. – Palaeontographica A **250**: 123-154.

EVANS, S. E. & BARBADILLO, L. J. (1997): Early Cretaceous lizards from Las Hoyas, Spain. – Zoological Journal of the Linnean Society **119**: 23-49.

HOFFSTETTER, R. (1967): Coup d'oeil sur les sauriens (= lacertiliens) des Couches de Purbeck (Jurassique supérieur d'Angleterre). – Colloques Internationales du Centre National de la Recherche Scientifique **163**: 349-371, Paris.

LANG, M. (1991): Generic relationships within Cordyliformes (Reptilia: Squamata). – Bulletin van het Koninklijk Belgisch Instituut voor Natuurwetenschappen **61**: 121-188.

PROTHERO, D. & ESTES, R. (1980): Late Jurassic lizards from Comio Bluff, Wyoming and their palaeobiographic significance. – Nature **286**: 484-486.

RICHTER, A. (1994): Lacertilia aus der Unteren Kreide von Uña und Galve (Spanien) und Anoual (Marokko). – Berliner geowissenschaftliche Abhandlungen E **14**: 1-147.

RIEPPEL, O. (1984): Miniaturization of the Lizard Skull: Its Functional and Evolutionary Implications. – In: FERGUSON, M. W. [ed.] The Structure, Development and Evolution of Reptiles. – Symposia of the Zoological Society of London **52**: 503-520.

SEIFFERT, J. (1973): Upper Jurassic lizards from Central Portugal. – Memórias dos Serviços geológicos de Portugal **22**: 7-85.

STARCK, D. (1979): Vergleichende Anatomie der Wirbeltiere auf evolutionsbiologischer Grundlage, 2: Das Skeletsystem. – 776 pp.; Berlin, Heidelberg (Springer).

The crocodiles from the Guimarota mine

BERNARD KREBS & DANIELA SCHWARZ

Crocodiles and birds are the sole survivors of a group called Archosauria, which also includes dinosaurs, pterosaurs and other, more primitive forms. Archosaurs are diapsid reptiles (reptiles with two pairs of openings between the eyes and the posterior end of the skull, the so-called temporal openings), which are distinguished from other members of this group by the presence of a preorbital opening between the narial opening and the eye opening, the orbit. The archosaurs are furthermore often characterized by a tendency for bipedal (two-legged) locomotion. This may be true for the dinosaurs, but not for the lineage that leads towards the crocodiles: the ancestors of the latter, the pseudosuchians, developed an advanced style of quadrupedal locomotion during the Triassic, in which the body was lifted off the ground and the legs moved parallel to the axis of symmetry of the body. A highly specialized tarsal joint appeared – similar to that of mammals – in which the movement mainly takes place between the astragalus and calcaneum. Apart from the specialisations in the locomotor apparatus, in crocodiles the skull becomes flattened and the snout more or less elongate, while the preorbital openings are being reduced. In the course of crocodile phylogeny, a secondary palate is developed – also in analogy to mammals – which separates the nasal passages from the mouth.

The first crocodiles appear in the Late Triassic. They are clearly terrestrial animals. In Jurassic sediments, however, mainly crocodiles that are adapted to a life in water are found. This is partially due to the fact that Europe, the best known area in the world in terms of paleontology, was largely covered by oceans during the Jurassic and Cretaceous, as noted in the introduction. These marine crocodiles are certainly not the ancestors of the modern crocodiles. Only little evidence of the history of the terrestrial or amphibious crocodiles is known from the Jurassic. Thus, with its ecosystem at the boundary of land and water, the Guimarota mine provides important information concerning this stage of crocodilian evolution.

First, however, a visitor to this environment will be introduced, the marine crocodile *Machimosaurus hugii*.

Machimosaurus hugii V. MEYER 1837

When W. G. KÜHNE and his helpers worked in Guimarota in 1961, the last year of commercial coal mining in the Guimarota mine, remains of a very large reptile were found. Since all fragments came from a small area and no bones of this size had otherwise ever been found, it was evident that all elements belonged to a single individual. It seems probable that the whole skeleton was originally present, but was not recognized by the miners underground.

The material comprises two larger fragments of the snout, remains of the frontal, the left postorbital, the left pterygoid and the cranial base, as well as two dorsal vertebrae, two sacral vertebrae, a caudal vertebra, ribs, the left fibula and several osteoderms. The latter exhibit a conspicuous pattern of elongate oval grooves on their outer surface. An elongated snout and a secondary palate can be recognized in the remains of the upper jaws, which – as was the case with the pattern of grooves on the osteoderms – is sufficient to identify the fossil as a crocodile. The teeth that are preserved in the jaws are bluntly conical and round in cross section. Their surface is covered by fine, irregular, often interrupted ridges, which terminate in a tangled lace at the tip. Two opposed, ornamented ridges indicate cutting edges (Fig. 10.1). Such teeth were known for a long time from the Late Jurassic of Solothurn (Switzerland) and had already been described in 1837 under the name of *Machimosaurus hugii* by H. V. MEYER, the founder of German vertebrate paleontology. Isolated *Machimosaurus* teeth were also found in other marine sediments of Late Jurassic age in Europe. However, since no determinable skeletal elements had ever been found in association with the teeth, the systematic position of *Machimosaurus* remained unknown. A slight similarity with the teeth of the geologically younger,

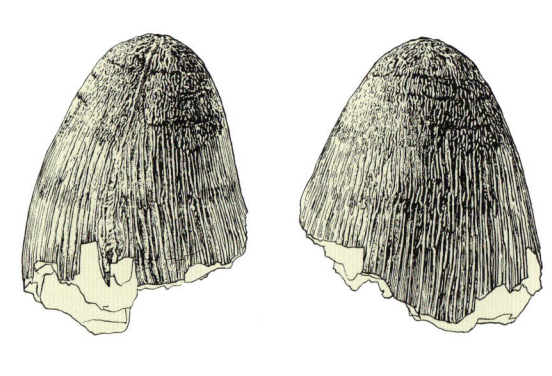

Fig. 10.1. *Machimosaurus*. A typical tooth in anterior (left) and external view (right). The tooth is bluntly conical, round in cross-section and exhibits a characteristic ornamentation. Height 2.5 cm.

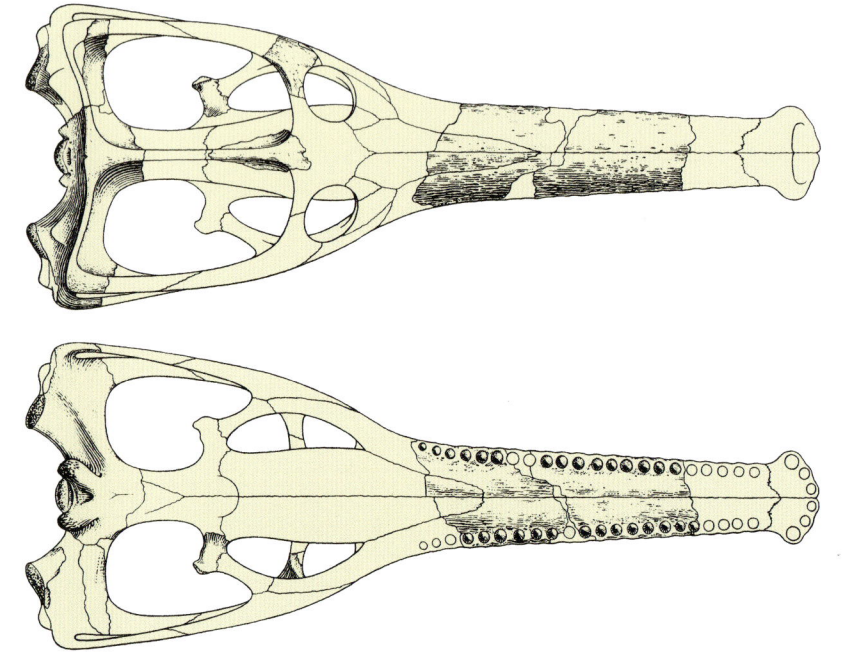

Fig. 10.2. *Machimosaurus*. Reconstruction of the skull in dorsal (upper) and ventral view (lower). Reconstructed (unknown) parts are shown in outline only. Note the elongate snout with many homogeneous teeth and the large upper temporal fenestrae. Length c. 130 cm.

terrestrial-amphibious crocodile *Goniopholis* (see below) led to the referral of *Machimosaurus* to the Goniopholodidae, despite the obvious contradictions in the age and environments.

For the first time, teeth of *Machimosaurus* were now found in association with determinable skeletal elements. The skull could be reconstructed on the basis of the preserved elements (KREBS 1967, 1968). It became clear that the animal had a strongly elongate snout and enormous upper temporal openings, which were only separated by a thin bridge of bone (Fig. 10.2). Thus, *Machimosaurus* could be identified as a representative of the family Teleosauridae, a group of the Mesozoic marine crocodiles, which are united in the Thalattosuchia. The subsequent find of a complete skull of *Machimosaurus* from the Late Jurassic of Montmerle (Département Ain, France) confirmed the reconstruction of the skull (BUFFETAUT 1982a). The reconstructed skull has a length of 1.5 m; the complete animal must have been more than 9 m long.

The adaptations for a life in the open ocean is also evident from a special character of the dorsal vertebrae: the articular surfaces of the zygapophyses, the processes with which the vertebrae articulate with each other, are very steeply inclined in *Machimosaurus*, as is the case in the other teleosaurids (KREBS 1962). Thus, lateral movements of the body, which are important for walking as well as for wriggling swimming movements, are strongly restricted. Such a stiffening of the body, however, is advantageous for fast swimming; the propulsion force is displaced posteriorly and generated by wriggling movements of the tail.

The long snout with its high number of homogeneous teeth forms a so-called fish rake, as it is found in many secondarily aquatic terrestrial vertebrates that feed on swimming prey, such as ichthyosaurs, or dolphins. The enormous upper temporal openings give testimony of the enlargement of a part of the jaw musculature, the Musculus pseudotemporalis, which enables an extremely rapid closing of the jaws. Below the occipital condyle of *Machimosaurus*, an extremely strongly developed tuber basioccipitalis is found on each side: these processes provided attachment areas for the cervical musculature (Fig. 10.2, lower). Among Recent crocodiles, such muscle attachment areas are only found in the aquatic, fish-eating gharial. They enable the animal to rapidly swing the head laterally to catch a prey item – opening the jaws in a frontal approach on the prey would considerably slow down the swimming speed (HUA 1997). In contrast to the pointed teeth of most teleosaurids, the bluntly rounded teeth of *Machimosaurus* indicate hard, but swimming prey. Thus, possible prey items includes marine turtles or large ganoid fishes (bony fishes with thick, bony scales), both faunal elements which are known from the coals of Guimarota.

The marine crocodile *Machimosaurus* represents an alien element in the ecosystem of Guimarota. It can be assumed that this animal came to the coast to lay its eggs, as it is the case in Recent marine turtles.

Goniopholis cf. *simus*

Only one almost complete skeleton of an amphibious crocodile has been found in the coals of the Guimarota mine. It comprises a skull with associated isolated lower jaw fragments, osteoderms, various vertebrae and ribs and some disarticulated bones, some of which represent elements of the pectoral and pelvic girdles. Based on their corresponding size and association in the sediment, the remains can be regarded as belonging to a single individual (SCHWARZ 1999).

When the sediment condensed, the skull was slightly distorted laterally and strongly compressed dorso-ventrally. This resulted in a damage of the rather delicate secondary palate. The braincase is completely destroyed; its elements have been pressed against the more massive skull roof and are thus no longer identifiable. The anterior part of the snout with both premaxillae and parts of the maxillae are broken off. However, these parts are preserved as three isolated fragments, which can be fitted to the preserved parts of the articulated snout. It is therefore possible to complete the skull anteriorly. Based on the length of the skull, the total length of the individual can be estimated (WERMUTH 1964): a skull length of 26 cm indicates a total length of approximately 2 m for the animal. The slightly brevirostrine (short-snouted) skull exhibits a rather broad snout and strongly developed sculpturing on its dorsal side, which becomes less conspicuous laterally. It consists of a pattern of small grooves, as it is typical for crocodiles. The snout is strongly festooned vertically and horizontally. The skull is generally similar to Recent relatively short-snouted taxa, such as the caiman *(Caiman crocodilus)* or the black caiman *(Melanosuchus niger)*.

Only fragments are preserved of the corresponding lower jaw. They are, however, sufficient to reconstruct the jaw, which was very massive. Its elements are also strongly sculptured.

The preserved elements allow a taxonomic determination of the specimen. Within crocodiles, the animal can be referred to the advanced crocodiles, the Neosuchia (BENTON & CLARK 1988), based on the presence of a secondary palate, the arrangement of the elongate teeth in two waves in both the upper and lower jaw, and the development of a narrow, posteriorly directed retroarticular process in the lower jaw. The broad and slightly brevirostrine snout, the constriction at the suture between the premaxilla and maxilla, the postorbital bar, which is depressed in relation to the other skull roof elements, the participation of the frontals in the anterior margin of the upper temporal openings and the size ratios of the upper temporal openings in relation to the orbital open-

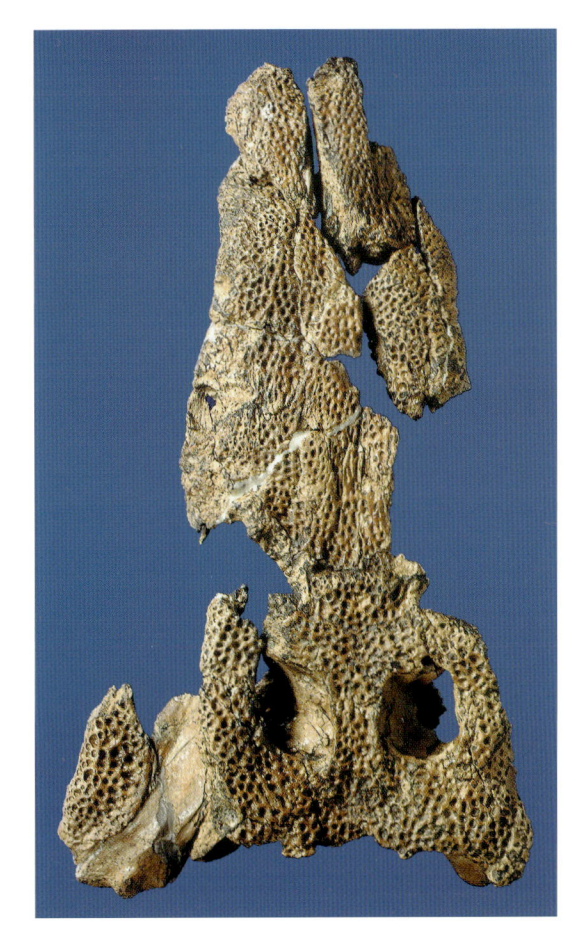

Fig. 10.3. Skull of *Goniopholis* cf. *simus* in dorsal view, supplemented by isolated fragments of the premaxilla and maxilla. The groove-pattern, which is typical for crocodiles, is especially well developed. The quadrangular upper temporal fenestrae and the constriction in the anterior part of the snout are well visible (Gui Croc 1/1-1/4). Length of the skull 24.2 cm.

Fig. 10.4. Fragment of a right lower jaw (Gui Croc 2) of another specimen of *Goniopholis* sp. indet. with four teeth in situ. The other teeth were lost before burial, since they were only weakly implanted in the jaw, due to their simple root. Length of the jaw fragment 12.3 cm.

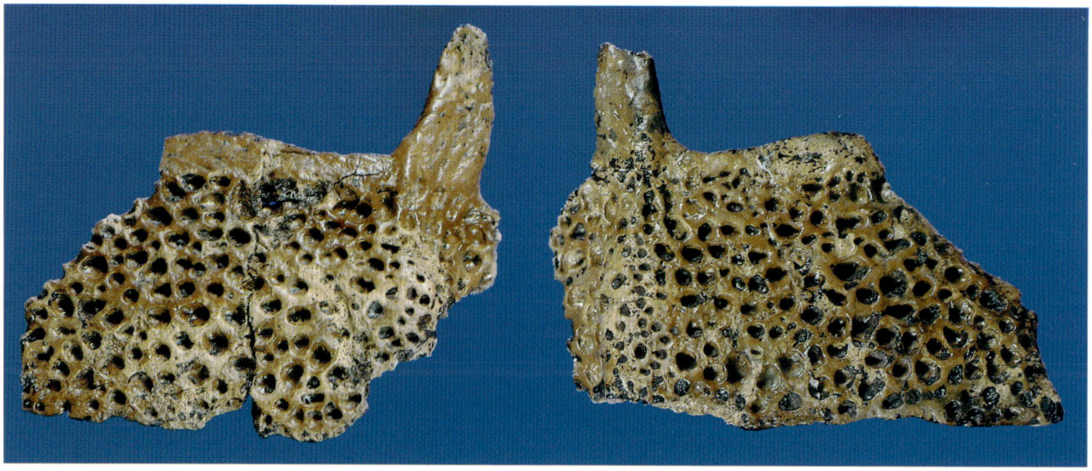

Fig. 10.5. Two dorsal osteoderms (Gui Croc 1/48a and b) from the anterior part of the body in dorsal view. The rectangular shape, the well-developed groove-pattern on the dorsal side and the anteriorly pointing peg-and socket joint are typical for *Goniopholis*. Width of the right plate 6.9 cm.

ing indicate that the specimen can be referred to the family Goniopholididae (characters according to STEEL 1973 and BUFFETAUT 1982b).

Within this family, the Portuguese remains belong to a representative of the type genus *Goniopholis*. The referral is possible because of the only preserved tooth, which exhibits characteristic vertical ridges and grooves in the enamel as well as two opposed cutting edges that are typical for the genus according to OWEN (1878) and STEEL (1973). Further characters of the genus are the osteoderms, which have already been described by OWEN (1878): the overlapping dorsal armor plates with a ball-and-socket joint and the hexagonal ventral armor plates which form a ventral shield, being connected to each other by bony sutures. Morphological details of the skull, such as the position of the external narial opening, the subquadrangular shape of the upper temporal openings, and a U-shaped, anteriorly flexed bony ridge on the frontal – as it is also found in the Recent broad-nosed caiman *(Caiman latirostris)* – have been described by OWEN (1878) and HOOLEY (1907) for the species *Goniopholis simus*. Thus, the specimen from Guimarota can probably referred to this species (the uncertainty in this referral is expressed by the cf. in the name). It is the geologically oldest representative of the genus *Goniopholis*.

Goniopholidids are known from the Late Jurassic of North America and western Europe and from the Cretaceous of western Europe, North America and eastern Asia. Their origin is still uncertain. In Europe, the genus *Goniopholis* is mainly found in the Early Cretaceous of England and the Iberian Peninsula, in fresh water deposits with or without brackish influence. These animals are generally similar to modern crocodiles and grow up to several meters in length. They moved through the water by wriggling movements of the body. The body was protected by dorsal osteoderms and a strongly developed ventral shield. *Goniopholis* stood at the top of the food chain in the aquatic environments of its time and fed on fishes and corpses of large terrestrial animals, such as dinosaurs. The animal found in the Guimarota mine probably lived close to its place of burial, where it found abundant food sources.

Lisboasaurus estesi

In his PhD thesis on the lizard material from the first excavations in the Guimarota mine, SEIFFERT (1970, published English version 1973) created the genus and species *Lisboasaurus estesi* for an unusual upper jaw fragment. One tooth is preserved in the approximately 2 cm long bone. Its transversely flattened, lanceolate shaped crown is placed on a widened basis. The outer surface of the jaw exhibits a conspicuous depression (Fig. 10.6). SEIFFERT (1973) referred some other, poorly preserved specimens to the same taxon and erected a second species, *Lisboasaurus mitracostatus* on the basis of a fragment of a lower jaw, which is difficult to interpret. He regarded *Lisboasaurus* as a lacertilian, related to monitor lizards.

In the course of a review of the material described by J. SEIFFERT, the English paleontologist S. E. EVANS interpreted the lateral depression as a preorbital opening, which characterizes this animal as an archosaur. However, together with A. R. MILNER, she went even further, in interpreting it as a bird ancestor and referring the genus to the "Maniraptora", a group that includes birds and those dinosaurs that are closely related to birds (MILNER & EVANS 1991). However, this interpretation did not convince the vertebrate paleontologists in Berlin.

The discovery of a complete skeleton of a small crocodile in the Early Cretaceous of Las Hoyas (province of Cuenca, Spain), which is identical with *Lisboasaurus* in the morphology of the upper jaw and the shape of the teeth, led to the final clarification of the systematic position of this genus: it is a crocodile. The *Lisboasaurus*-skeleton from Las Hoyas demonstrates that it is a small, on the basis of its preorbital opening rather basal crocodile, which was long-limbed and digitigrade (only the toes are set on the ground) and thus a strictly terrestrial animal (BUSCALIONI & ORTEGA 1995; BUSCALIONI et al. 1996).

Further material of crocodiles

Apart from the remains described above, many isolated crocodilian bones and innumerable crocodile teeth have been found in the examination of the coal and the picking of the concentrate resulting from screen washing.

The identification of isolated bones is difficult and requires further studies of the material. The size of the remains ranges from 3 mm to 2 cm. The material mainly consists of elements of the skull and lower jaw, although postcranial bones and small osteoderms are also present. Significantly larger specimens are sometimes also found, such as a 5 cm long vertebra that can be referred to a goniopholidid like the one described above.

Isolated teeth are extraordinarily common. Many of these can be referred to the genus *Goniopholis*: they are relatively stout, blunt, slightly curved and show characteristic flutings in the enamel (see above). The largest of these teeth is 1.5 cm high; it represents a large individual of more than 2 m total length.

A high number of knob-like teeth are also present, which are only few millimeters large. They represent durophagous (i.e. adapted for hard-shelled food items) forms, such as *Bernissartia*. Small and slightly larger teeth with smooth enamel indicate the presence of a *Theriosuchus*-like genus. *Bernissartia* and *Theriosuchus* are advanced Mesozoic crocodilians that belong to the Neosuchia. The animals were less than one meter long. While the *Bernissartia*-like animals preferred clams and snails, the *Theriosuchus*-like forms were small carnivores, which fed on lizards and mammals; perhaps they were also specialized for insects (BRINKMANN 1989). The small isolated bones that were mentioned above might be referable to these crocodiles.

The extremely high abundance of isolated crocodile teeth is at least partially due to the continuous tooth replacement in crocodiles. Thus, a disproportional percentage of teeth in comparison with skeletal material is often found in fossil localities, and proof of the presence of rare species is often provided by teeth in fossil lagerstatten.

The especially high abundance of *Goniopholis* teeth indicates that this genus was common in the area were the coals of Guimarota were deposited. The smaller *Theriosuchus*- and *Bernissartia*-like forms probably rather lived in the further surroundings, in small rivers and lakes. They preferred the hinterland, without any marine influence. Their remains were transported by rivers over short distances into the depositional environment of Guimarota.

Fig. 10.6. Right maxilla of *Lisboasaurus estesi* ("Gui 37", holotype) in lateral view, showing the depression (top left) that was interpreted as a preorbital fenestra by MILNER & EVANS (1991). Length 1.9 cm.

References

BENTON, M. J. & CLARK, J. M. (1988): Archosaur phylogeny and the relationship of the Crocodylia. – In: BENTON, M. J. [ed.] The phylogeny and classification of the tetrapods **1**: 295-338, Oxford (Clarendon Press).

BRINKMANN, W. (1989): Vorläufige Mitteilung über die Krokodilier-Faunen aus dem Ober-Jura (Kimmeridgium) der Kohlengrube Guimarota, bei Leiria (Portugal) und der Unter-Kreide (Barremium) von Uña (Provinz Cuenca, Spanien). – Documenta naturae **56**: 1-28.

BUFFETAUT, E. (1982a): Le crocodilien *Machimosaurus* VON MEYER (Mesosuchia, Teleosauridae) dans le Kimméridgien de l'Ain. – Bulletin trimestriel de la Société géologique de Normandie **69**: 17-27.

– (1982b): Radiation évolutive, paléoécologie et biogéographie des crocodiliens mésosuchiens. – Mémoires de la Société géologique de France (nouvelle Sér.) **142**: 1-88.

BUSCALIONI, A. D. & ORTEGA, F. (1995): Crocodylomorphs. – In: II international Symposium on lithographic Limestones, Lleida-Cuenca (Spain) 1995, Las Hoyas field trip guide book: 59-61, Madrid (Universidad Autónoma).

BUSCALIONI, A. D.; ORTEGA, F.; PÉREZ-MORENO, B. P. & EVANS, S. E. (1996): The Upper Jurassic maniraptoran theropod *Lisboasaurus estesi* (Guimarota, Portugal) reinterpreted as a crocodylomorph. – Journal of Vertebrate Paleontology **16**: 358-362.

HOOLEY, R. W. (1907): On the skull and greater portions of the skeleton of *Goniopholis crassidens* from the Wealden Shales of Atherfield (Isle of Wight). – Quarterly Journal of the geological Society of London **63**: 50-63.

HUA, S. (1997): Adaptations des crocodiliens mésosuchiens au milieu marin. – Mémoires des Sciences de la Terre, Université Pierre et Marie Curie, Paris **97**: 1-209.

KREBS, B. (1962): Ein *Steneosaurus*-Rest aus dem Oberen Jura von Dielsdorf, Kt. Zürich, Schweiz. – Schweizerische paläontologische Abhandlungen **79**: 1-28.

– (1967): Der Jura-Krokodilier *Machimosaurus* H. V. MEYER. – Paläontologische Zeitschrift **41**: 46-59.

– (1968): Le crocodilien *Machimosaurus*. – Memórias dos Serviços geológicos de Portugal (nova Sér.) **14**: 21-53.

MEYER, H. VON (1837): [Mittheilungen, an Professor BRONN gerichtet]. – Neues Jahrbuch für Mineralogie, Geognosie, Geologie und Petrefaktenkunde **1837**: 557-562.

MILNER, A. R. & EVANS, S. E. (1991): The Upper Jurassic diapsid *Lisboasaurus estesi* – a maniraptoran theropod. – Palaeontology **34**: 503-513.

OWEN, R. (1878): Monograph on the fossil Reptilia of the Wealden and Purbeck formations, Supplement VIII: Crocodila *(Goniopholis*, *Petrosuchus*, and *Suchosaurus)*: 1-15, London (Palaeontographical Society).

SCHWARZ, D. (1999): Das Skelett eines Goniopholididen (Crocodilia) aus dem Oberen Jura von Portugal. – unpublished Diploma thesis, Fachbereich Geowissenschaften der Freien Universität Berlin, 108 pp, Berlin.

SEIFFERT, J. (1970): Oberjurassische Lazertilier aus der Kohlengrube Guimarota bei Leiria (Mittelportugal). – Unpublished PhD thesis, Mathematisch-Naturwissenschaftliche Fakultät der Freien Universität Berlin, 180 pp., Berlin.

– (1973): Upper Jurassic lizards from Central Portugal. – Memórias dos Serviços geológicos de Portugal, (nova Sér.) **22**: 7-85.

STEEL, R. (1973): Crocodylia. In: KUHN, O. [ed.] Handbook of Palaeoherpetology **16**: 1-116, Stuttgart (Gustav Fischer).

WERMUTH, H. (1964): Das Verhältnis zwischen Kopf-, Rumpf- und Schwanzlänge bei rezenten Krokodilen. – Senckenbergiana biologica **45**: 369-385.

The dinosaur fauna from the Guimarota mine

OLIVER W. M. RAUHUT

Dinosaur remains are quite common in the Guimarota mine, as was expected in a Jurassic locality. More than 600 isolated dinosaur teeth could be identified so far; in contrast, jaws and other bones are extremely rare. Although dinosaur teeth are not as diagnostic as those of mammals, they can still provide interesting information on the Late Jurassic fauna from Portugal. One observation concerning the dinosaur remains from Guimarota is especially noteworthy: while the name „Dinosauria" is usually associated with enormous giants, such as *Brachiosaurus*, the remains from Guimarota almost exclusively belong to small representatives of this group, to animals of less than two or even one meter total length. This peculiarity will be discussed in more detail after the systematic description of the remains.

Ornithischia

The exclusively herbivorous ornithischians ("bird-hipped dinosaurs") are only represented by the Ornithopoda ("bird-foot dinosaurs") in the Guimarota mine. The presence of other groups that are known from the Late Jurassic of Europe (WEISHAMPEL 1990), like the armored stegosaurs and ankylosaurs (which are united in the Thyreophora), could not be demonstrated so far.

In 1973, R. A. THULBORN described a new species of ornithopods, *Phyllodon henkeli*, on the basis of isolated teeth from the old excavation. The generic name refers to the leaf-like teeth of the animal (Figs. 11.1, 11.2), whereas the specific name was given in honor of Prof. S. HENKELs.

Phyllodon is a small representative of the Hypsilophodontidae, a group of generally rather small, bipedal ornithopods that had an almost worldwide distribution from the Middle Jurassic to the Late Cretaceous (WEISHAMPEL 1990). Hypsilophodontids are basal representatives of a group of ornithopods that was especially successful in the Cretaceous. They are the first group to exhibit a character, which probably was essential for the great success of later ornithopods: the development of an effective chewing mechanism. Similar to the Recent ungulates, a transverse movement of the jaws was employed. However, whereas the latter achieve this transverse motion by lateral movements of the lower jaw and only chew on one side of the mouth, in ornithopods the jaw musculature generated a pressure on the upper dentition via the lower dentition when the teeth were in occlusion, which forced the kinetic upper jaws outwards. When the pressure was released, tendons in the skull moved the upper jaws back into their original position (WEISHAMPEL 1984). Outside the Mammalia, such an effective chewing mechanism is unique, and it probably represented a great advantage of the ornithopods over other herbivores.

The identification of dinosaur species on the basis of isolated teeth is difficult, and thus, the systematic position, as well as the taxonomic validity of *Phyllodon* were long uncertain. Whereas GALTON (1983) considered the species to be valid and even provided a revised diagnosis for it, it was listed as a nomen dubium by SUES & NORMAN (1990) in their review of the Hypsilophodontidae. However, new material from the systematic dig from 1973 to 1982, including some jaw fragments, as well as new information on the taxonomic utility of dental characters in at least some groups of ornithischians (BAKKER et al. 1990, THULBORN 1992) led to the conclusion that *Phyllodon henkeli* represents a valid species (RAUHUT in prep.). The closest relatives of this taxon are probably the more or less contemporaneous species *Othnielia rex* and *Drinker nisti* from the Late Jurassic Morrison Formation of North America (GALTON 1983, BAKKER et al. 1990).

Another group of ornithopods, the iguanodontians, are only known from three teeth from Guimarota (Fig. 11.3). Unfortunately, the teeth of these animals are not as diagnostic as those of the hypsilophodontids, so that they cannot be referred to any genus or species. They probably represent an animal like *Camptosaurus*, a basal iguanodontian, which is especially well known from the Morrison Formation of North America, but is also present in the Late Jurassic of Europe (GALTON & POWELL 1980). The teeth from the Gui-

Fig. 11.1. Lateral teeth of *Phyllodon henkeli* THULBORN 1973; **a** maxillary tooth Gui Or 10 in lingual view, **b** dentary tooth Gui Or 13 in labial view. Scale bar = 1 mm.

Fig. 11.2. Premaxillary tooth of *Phyllodon henkeli* in lingual view (Gui Or 24). Height 4 mm. ▷

▷▷
Fig. 11.4. Teeth of a ?brachiosaurid sauropod; left: Gui Sd 1 in lingual view, right: Gui Sd 3 in labial view. Height of the left tooth 2.6 cm.

Fig. 11.3. Maxillary tooth of an undetermined iguanodontian in labial view (Gui Or 5). Scale bar = 1 mm.

marota mine represent one of the oldest occurrences of this group. Being the sister group of the hypsilophodontids, they were equipped with the same chewing mechanism. The great majority of the Cretaceous ornithopods belong to the Iguanodontia, including the well known hadrosaurids ("duck-billed dinosaurs").

Saurischia

The Saurischia ("lizard-hipped dinosaurs") are represented in Guimarota by both of their major groups, the Sauropoda and the Theropoda. Theropod teeth are the most abundant dinosaur fossils in the Guimarota mine.

Sauropoda

Although they are one of the most common dinosaur groups of the Late Jurassic world wide (WEISHAMPEL 1990), the sauropods ("brontosaurs") are only represented by five teeth and tooth fragments in the Guimarota mine (Fig. 11.4). Based on their slender, but slightly spatulate shape, they can probably be referred to a representative of the Brachiosauridae. Since the genus *Brachiosaurus* has already been described from the Late Jurassic of Portugal (LAPPARENT & ZBYSZEWSKI 1957), it seems possible that the teeth from Guimarota also represent this genus; however, sauropod teeth are not diagnostic on generic or specific level.

Brachiosaurids are known since the Middle Jurassic and are found in Europe from the early Late Jurassic to the higher Early Cretaceous (MCINTOSH 1990, WEISHAMPEL 1990). It is unknown if they survived until the Late Cretaceous. They are mainly distinguished from most other sauropods by their relatively long forelimbs, which are longer than the hindlimbs in at least some representatives, their relatively short tail, and the broad, high skull with extremely large external narial openings.

Sauropods are the giants of the Mesozoic. Even if one speaks about small sauropods, this usually still implies animals of more than 10 m in length. Especially the brachiosaurids represent some of the largest known terrestrial vertebrates (JANENSCH 1950, COLBERT 1962). It is thus very surprising that even the teeth of the brachiosaurids from the Guimarota mine indicate rather small animals; the height of the most complete tooth crown from Guimarota is 2.5 cm, and thus only half the size of teeth of a comparable position within the mouth in the smallest known skull of *Brachiosaurus brancai* from the Late Jurassic of Tendaguru (JANENSCH 1935-36). Unfortunately, it cannot be determined whether the taxon from the Guimarota mine represents a small taxon of brachiosaurids, or a juvenile of a larger taxon, on the basis of teeth alone. Sauropods were also very effective herbivores. However, they probably processed their food with the help of gizzard stones, rather than with their teeth (DODSON 1991).

Whereas it was formerly generally assumed that sauropods could only support their enormous weights if they were at least partially submerged in water, convincing arguments have recently been brought forward for an almost exclusively terrestrial lifestyle of these animals (COOMBS 1975, DODSON 1991). Thus, the swampy coastal plain in which the coals of Guimarota were deposited does probably not represent the main habitat of the brachiosaurid from this locality; this might be one reason for the scarcity of sauropod remains in the Guimarota mine.

Theropoda

As mentioned above, theropod teeth are the most abundant dinosaur remains from the Guimarota mine. However, not only the great abundance of fossils, but especially the taxonomic composition of the theropod fauna from Guimarota is highly interesting (ZINKE & RAUHUT 1994, WEIGERT 1995, ZINKE 1998). Unfortunately, other skeletal elements are extremely rare and can usually not be determined with certainty (Fig. 11.5).

Many teeth can be referred to the genus *Compsognathus* (Fig. 11.6). This taxon was first described on the basis of an almost complete skeleton from the famous Late Jurassic lithographic limestones of Solnhofen (WAGNER 1861) and is meanwhile also known from the approximately contemporaneous plattenkalks of Canjuers (southern France; BIDAR et al. 1972). Only recently, a close relative was discovered in the Early Cretaceous of China (CHEN et al. 1998). *Compsognathus* and its relatives are some of the smallest known dinosaurs: the holotype of the genus is only 60 cm long,

Fig. 11.5. Claw of the hand of a theropod (Gui Th 5). Length 1.7 cm.

and the teeth from Guimarota also indicate small animals. Another aspect worth mentioning in this taxon is the high degree of heterodonty in its jaws (Fig. 11.7; STROMER 1934, OSTROM 1978). The teeth in the anteriormost parts of the jaws are large, pointed, strongly recurved and have an almost round cross-section at their basis. In the middle of the jaws, the teeth exhibit a rather typical theropod tooth morphology: they are transversely flattened, recurved and have serrated cutting edges. The posteriormost teeth are generally similar, but considerably lower and relatively broader. All of these different types of teeth were found in the material from Guimarota (ZINKE 1998). Both the holotype of *Compsognathus*, and a specimen of the Chinese compsognathid *Sinosauropteryx* have the remains of their last meal preserved in their stomach region (OSTROM 1978, CHEN et al. 1998); in both cases, they are the remains of a lizard. Thus, the compsognathid from the Guimarota mine probably also preyed on the smaller tetrapods known from this locality, including mammals and lizards.

Several larger teeth are similar in their morphology to teeth of allosaurs and similar basal tetanurans and thus probably indicate the pres-

Fig. 11.6. Life reconstruction of the small theropod *Compsognathus longipes* WAGNER 1861. This, or a closely related species was also present in the Guimarota ecosystem. The reconstruction is based on the holotype of *C. longipes* (Bayerische Staatssammlung für Paläontologie und historische Geologie AS. I. 563). The hair-like covering of the body is based on preserved integumentary structures in a close relative from China (*Sinosauropteryx*; CHEN et al. 1998). Of course, these structures are not homologous to the hairs of mammals, but probably represent proto-feathers. The camouflage pattern is pure speculation; similar patterns are found in Recent lacertilians (e.g. *Chamaeleo*). Length of the animal approximately 60 cm.

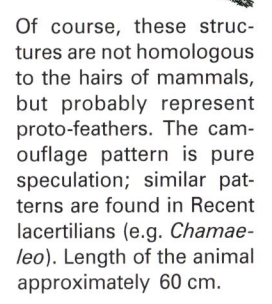

Fig. 11.7. Dentition of *Compsognathus longipes*. Note the high degree of heterodonty, which is unusual for a reptile; **a** premaxillary tooth, **b** maxillary tooth from the middle of the tooth row, **c** posterior maxillary tooth, **d** anterior dentary tooth, **e** dentary tooth from the middle of the tooth row, **f** posterior dentary tooth. After STROMER (1934) and the holotype of *C. longipes*. Approximately ten times original size.

Fig. 11.8. Teeth of larger theropods from the Guimarota mine; **a** lateral tooth of an *Allosaurus*-like theropod in labial view (Gui D 66), **b** tooth of a *Ceratosaurus*-like theropod in lingual view (Gui D 191). Height 3.3 cm (a) and 1.8 cm (b).

Fig. 11.9. Tooth of cf. *Richardoestesia* (Gui D 122). Height 5 mm.

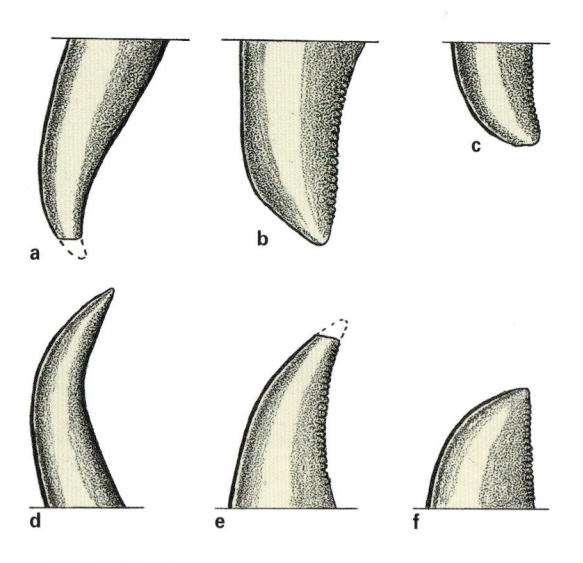

ence of a representative of such a group of theropods (Fig. 11.8a). These groups are well known from the Jurassic (MOLNAR et al. 1990, WEISHAMPEL 1990), and remains of them are also found in other Late Jurassic localities in Portugal (LAPPARENT & ZBYSZEWSKI 1957, RAUHUT & KRIWET 1994) Therefore, their presence in the Guimarota mine is not surprising.

A further taxon, which is especially well known from the Late Jurassic of North America, the genus *Ceratosaurus*, probably also had a European relative in the Guimarota mine; several teeth from this locality have great similarities with the rather characteristic teeth of this genus (Fig. 11.8b).

Apart from these typical Jurassic theropods, the fauna from Guimarota contains several advanced groups, which are otherwise mainly known from the Cretaceous. A taxon of uncertain systematic position is represented by many teeth and a jaw fragment. While the jaw is rather similar to that of *Compsognathus*, the teeth show several advanced characters, such as a basal constriction between crown and root and certain details in the serrations. Based on this combination of characters, this jaw fragment probably represents a relative of the mainly Cretaceous coelurosaurs (ZINKE & RAUHUT 1994). A conspicuous character of the teeth are well developed longitudinal ridges on both sides of the crown. These ridges exhibit a typical pattern in all of the teeth preserved and are thus probably a taxonomic character, rather than a pathology (CURRIE et al. 1990, ZINKE & RAUHUT 1994). However, if the taxon from the Guimarota mine is really closely related to the Late Cretaceous North American theropod *Paronychodon*, as was assumed by ZINKE & RAUHUT (1994) and ZINKE (1998), seems questionable, since several significant differences in the morphology of the teeth are found.

A further theropod of uncertain systematic position is represented by slender, dagger-like teeth with very small serrations (Fig. 11.9). The teeth are virtually identical to those of the Late Cretaceous North American genus *Richardoestesia* (CURRIE et al. 1990) und thus probably represent a closely related taxon (ZINKE 1998). Unfortunately, not enough material of *Richardoestesia* is known to determine its systematic position. However, the presence of very similar teeth in the Late Jurassic of Portugal and the Early Cretaceous of Spain (RAUHUT & ZINKE 1995) indicate that this genus represents a long-lived, though obviously still poorly known group of theropods.

A few teeth from the Guimarota mine can be referred to the Dromaeosauridae (ZINKE 1998). This group, which became famous due to the appearance of its representative *Velociraptor* in the movie "Jurassic Park", is otherwise also almost exclusively known from the Cretaceous. Both subfamilies of the Dromaeosauridae are present in Guimarota, based on their characteristic teeth. The teeth of the dromaeosaurines are characterized by a conspicuous inward flexure of the anterior cutting edge (Fig. 11.10; CURRIE et al. 1990, CURRIE 1995), while those of the velociraptorines exhibit a significant difference in the size of the denticles on the anterior and posterior cutting edges (Fig. 11.11; OSTROM 1969, CURRIE et al. 1990).

Dromaeosaurids are interesting for two main reasons: there is evidence that at least some representatives of this group were pack hunters (OSTROM 1970, MAXWELL & OSTROM 1995), and dromaeosaurids are usually regarded as the sister group to birds (OSTROM 1976, HOLTZ 1994). If the proposed pack-hunting behavior was also present in the dromaeosaurids from Guimarota can of course not be determined on the basis of isolated teeth. However, the sister group relationship with birds are of great interest, since one of the main arguments of the critics of the theory of the theropod origin of birds has always been that dromaeosaurids appear much later in the fossil record than the oldest known bird, *Archaeopteryx*, from the Late Jurassic (Tithonian) lithographic limestones of Bavaria (e.g., MARTIN 1983). This argument can be invalidated with the find of dromaeosaurids in the Late Jurassic, contemporaneous with *Archaeopteryx*. This is a good example, how even fragmentary material might sometimes help in solving important problems.

A further type of tooth from the Guimarota mine indicates the presence of another group that is almost exclusively known from the Cretaceous, the troodontids (Fig. 11.12). Teeth of troodontids are distinguished by their relatively large serrations with typical rounded grooves at the basis of each denticle, their lenticulate cross-section and the well-developed constriction between crown and root (CURRIE 1987, CURRIE et al. 1990). Troodontids are closely related to dromaeosaurids and share with the latter the derived character of an enlarged claw on the second digit of the foot. It is furthermore worth mentioning that this group exhibits the largest relative brain volume amongst non-avian dinosaurs: in relation to the body mass, the brain of *Troodon formosus* is approximately 5.8 times larger than that of a crocodile (HOPSON 1980). As is the case for the other theropod remains, the teeth of the troodontid from Guimarota indicate rather small animals.

The genus *Stokesosaurus* is the only taxon of theropods that is represented by identifiable skeletal material in the Guimarota mine. A small right ilium, which is only 8.5 cm long (Fig. 11.13), can be referred to this genus, which was originally also described on the basis of an isolated ilium from the Late Jurassic Morrison Formation of North America (MADSEN 1974). The ilium of *Stokesosaurus* is characterized by a well-developed, sharply defined vertical ridge over the acetabulum and the extremely broad and short pubic peduncle (Fig. 11.13). The specimen from Guimarota differs from the type specimen of *Stokesosaurus clevelandi* from the Morrison Formation, which is almost twice as large, mainly in the proportions; however, it cannot be determined at present, whether these differences indicate the presence of a different species, or if they are only due to ontogenetic variation. Although known only from fragmentary material, the genus *Stokesosaurus* is also of great interest. A vertical ridge above the acetabulum, as it is found in this taxon, is otherwise almost exclusively known from tyrannosaurids (MOLNAR et al. 1990); thus, *Stokesosaurus* has repeatedly been interpreted as the oldest known possible tyrannosaurid (MADSEN 1974, BRITT 1991). This theory is now supported by the presence of several tyrannosaurid-like teeth in the Guimarota mine, including two of the diagnostic, stout premaxillary teeth, which are D-shaped in cross-section (Fig. 11.14; ZINKE 1998). Together with the discovery of a tyrannosaur-like braincase in the Morrison Formation (CHURE & MADSEN 1998) the material from Guimarota thus indicates that the origin of tyrannosaurids also reaches back to at least the early Late Jurassic. Furthermore, it seems that the first tyrannosaurids were rather small animals, and only their Late Cretaceous representatives grew to giant size,

Fig. 11.10. Tooth of a dromaeosaurine dromaeosaurid in lingual view (Gui D 77). Scale bar = 2 mm.

Fig. 11.11. Tooth of a velociraptorine dromaeosaurid (Gui D 67). Height 3 mm.

Fig. 11.12. Tooth of a troodontid with preserved root (Gui D 98). Scale bar = 1 mm.

Fig. 11.13. Right ilium of *Stokesosaurus* sp. in lateral view (Gui Th 1). Length 8.5 cm.

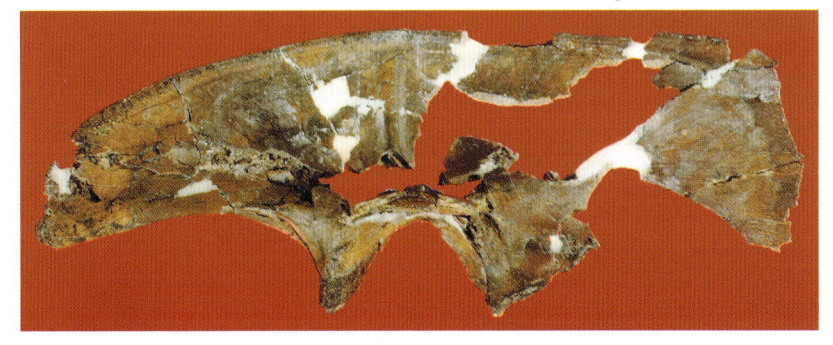

Fig. 11.14. Premaxillary tooth of a tyrannosaurid (Gui D 91), possibly *Stokesosaurus*; **a** mesial view, **b** lingual view. Scale bar = 2 mm.

such as the well known *Tyrannosaurus rex*. This theory is in general accordance with the assumption that the tyrannosaurids belong to the generally rather small coelurosaurs (HOLTZ 1994).

Another advanced theropod is present in the Guimarota mine. This taxon, however, is a well known element of the Late Jurassic theropod fauna. In 1995, A. WEIGERT described several teeth from the Guimarota mine which are very similar to those of the basal bird *Archaeopteryx*. For the first time, these specimens allowed the study of archaeopterygiform teeth in three dimensions, and these studies resulted in some interesting findings. The teeth are characterized by their sigmoidal shape and the inturned cutting edges, and exhibit rudimentary serrations on the anterior cutting edge, which are only visible under the scanning electron microscope (WEIGERT 1995, WIECHMANN & GLOY 2000*). Therefore, the Archaeopterygiformes were more primitive in their tooth morphology than later birds und thus represent a transformational model between theropods and birds in this respect. On the other hand, the teeth are strongly modified and thus indicate a high degree of specialization in the feeding apparatus. Unfortunately, however, it is still unknown what this specialization was.

The significance of the theropod fauna

Considering the list of theropod taxa known from the Guimarota mine, one might get the impression that Portugal was a center of theropod evolution in the Late Jurassic. This is, however, most probably not the case; the presence of such an unusually diverse association can be explained by the taphonomy of the locality and the methods of exploration employed. Usually, small vertebrate remains are less likely to be preserved than larger specimens, and bones and teeth are often destroyed by humin acids in the soil (CARPENTER 1982). The special preservational circumstances in the Guimarota mine (see GLOY 2000*) prevented such a destruction and made the discovery of large amounts of microvertebrate remains possible. Furthermore, the exploration methods were especially devised for the discovery of small vertebrate remains. Since most of the advanced theropods seem to have been small animals, the material from Guimarota thus allows an evaluation of the high diversity of Late Jurassic theropod faunas. New finds of microvertebrates from the Morrison Formation of North America (CHURE 1995) show that the taxonomic composition is not necessarily exceptional for this age. Nevertheless, the taxonomic diversity of the fauna of Guimarota is noteworthy. Apart from typical Jurassic forms, many taxa are present, which are then widely distributed and highly successful in the Cretaceous. It is interesting to note that the presence of many of these groups was to be expected, based on their sister group relationships with other taxa (e.g. SERENO 1997). Thus, microvertebrate localities like Guimarota can provide important information on the recognition of gaps in the fossil record, and can sometimes even help filling these gaps.

Taphonomy and ecology of the dinosaurs from Guimarota

As noted in the introduction, large dinosaurs are missing completely in the Guimarota mine. Although this observation is especially striking in the sauropods, which are widely distributed and usually gigantic in the Late Jurassic, it is nevertheless also true for the ornithischians and theropods. In the latter, for example, the otherwise common basal tetanurans are only represented by a few teeth. One of the reasons for the absence of large animals might be that large, heavy animals probably avoided the soft substrate of the coaly swamp. The probably dense vegetation surrounding the locality might have been another obstacle for large animals. A comparison with other Late Jurassic European localities (LAPPARENT & ZBYSZEWSKI 1957, WEISHAMPEL 1990) reveals that the bias towards small forms in Guimarota is quite significant. Many taxa, which are known from other localities in Portugal, such as the stegosaurs, ankylosaurs, *Camptosaurus*, and several taxa of sauropods (see WEISHAMPEL 1990), are extremely rare or missing completely.

Whether the small dinosaurs from the Guimarota mine represent juveniles, or taxa that are also small as adults, is difficult to determine. However, even if mainly juveniles are represented, the extremely small size of most remains, as well as the fact that mainly groups are present that are known to be rather small, indicate that most of the taxa represented are rather small animals by dinosaurian standards. It could furthermore well be that small dinosaurs are much more common than it is generally assumed and that their apparent scarcity is due to taphonomic factors; there is some evidence that some taxa of dinosaurs were really small as adults (CALLISON & QUIMBY 1984).

A further observation in the dinosaur material of Guimarota is noteworthy: theropods represent the most abundant group of dinosaurs and account for approximately 90 % of the teeth. This is especially worth mentioning, since predators are usually much less common than herbivorous animals. However, the quantitative distribution of teeth does of course not directly reflect the real abundance of the animals. Many different factors influence the percentile distribution of isolated teeth, such as the total number of teeth within a single jaw, tooth replacement rates, and differences in preservation potential. Especially the latter probably represents a significant bias in the ratio between teeth of ornithischians and theropods. The effective chewing mechanism in ornithopods resulted in significant wear of the teeth, so that most of them were considerably worn down when they were finally replaced. This wear and thus the exposure of the relatively soft dentine made these teeth much more vulnerable to destruction than the teeth of theropods, which were protected by a hard layer of enamel. Furthermore, the strong wear of ornithischian teeth makes their identification more difficult, so that many specimens might have been overlooked in the concentrate resulting from the screen washing. However, even if all of these factors are taken into consideration, theropods are probably still overrepresented in the Guimarota mine. Unfortunately, the reasons for this are difficult to determine, given the material at hand. It might be possible that ornithischians avoided the swampy environment because of the high risk of predation by theropods and crocodiles, while theropods found abundant food sources in the habitat with its common small tetrapods and probably also insects.

References

BAKKER, R. T., GALTON, P., SIEGWARTH, J. & FILLA, J. (1990): A new latest Jurassic vertebrate fauna, from the highest levels of the Morrison Formation at Como Bluff, Wyoming. Part IV. The dinosaurs: a new *Othnielia*-like hypsilophodontid. – Hunteria **2**: 8-19.

BIDAR, A., DEMAY, L. & THOMEL, G. (1972): *Compsognathus corallestris*, nouvelle espèce de dinosaurien théropode du Portlandien de Canjuers (Sud-est de la France). – Annales du Musée d'Histoire Naturelle de Nice **1**: 9-40.

BRITT, B. B. (1991): Theropods of Dry Mesa Quarry (Morrison Formation, Late Jurassic), Colorado, with emphasis on the osteology of *Torvosaurus tanneri*. – Brigham Young University Geology Studies **37**:- 72.

CALLISON, G. & QUIMBY, H. M. (1984): Tiny dinosaurs: are they fully grown? – Journal of Vertebrate Paleontology **3**: 200-209.

CARPENTER, K. (1982): Baby dinosaurs from the Late Cretaceous Lance and Hell Creek formations and a description of a new species of theropod. – Contribution to Geology, University of Wyoming **20**: 23-134.

CHEN, P.-J., DONG, Z.-M. & ZHEN, S.-N. (1998): An exceptionally well-preserved theropod dinosaur from the Yixian Formation of China. – Nature **391**: 147-152.

CHURE, D. J. (1995): The teeth of small theropods from the Morrison Formation (Upper Jurassic: Kimmeridgian), Ut. – Journal of Vertebrate Paleontology **15** (suppl.): 23A.

CHURE, D. J. & MADSEN, J. H. J. (1998): An unusual braincase (?*Stokesosaurus clevelandi*) from the Cleveland-Lloyd Dinosaur Quarry, Utah (Morrison Formation, Late Jurassic. – Journal of Vertebrate Paleontology **18**: 15-25.

COLBERT, E. H. (1962): The weights of dinosaurs. – American Museum Novitates **2076**: 1-16.

COOMBS, W. P. J. (1975): Sauropod habits and habitats. – Palaeogeography, Palaeoclimatology, Palaeoecology **17**: 1-33.

CURRIE, P. J. (1987): Bird-like characteristics of the jaws and teeth of troodontid theropods (Dinosauria, Saurischia). – Journal of Vertebrate Paleontology **7**: 72-81.

– (1995): New information on the anatomy and relationships of *Dromaeosaurus albertensis* (Dinosauria: Theropoda). – Journal of Vertebrate Paleontology **15**: 576-591.

CURRIE, P. J., RIGBY, J. K. & SLOAN, R. E. (1990): Theropod teeth from the Judith River Formation of southern Alberta, Canada: 107-125. – In: CARPENTER, K. & CURRIE, P. J. [Eds.] Dinosaur systematics. Approaches and perspectives, Cambridge (Cambridge University Press).

DODSON, P. (1991): Life styles of the huge and famous. – Natural History **1991**: 34.

GALTON, P. M. (1983): The cranial anatomy of *Dryosaurus*, a hypsilophodontid dinosaur from the Upper Jurassic of North America and East Africa, with a review of hypsilophodontids from the Upper Jurassic of North America. – Geologica et Palaeontologica **17**: 207-243.

GALTON, P. M. & POWELL, H. P. (1980): The ornithischian dinosaur *Camptosaurus prestwichii* from the Upper Jurassic of England. – Palaeontology **23**: 411-443.

GLOY, U. (2000*): Taphonomy of the fossil lagerstatte Guimarota. – In: MARTIN, T. & KREBS, B. [eds.] Guimarota – a Jurassic ecosystem: 129-135, München (Verlag Dr. F. Pfeil).

HOLTZ, T. R. J. (1994): The phylogenetic position of the Tyrannosauridae: implications for theropod systematics. – Journal of Paleontology **68**: 1100-1117.

HOPSON, J. A. (1980): Relative brain size in dinosaurs. Implications for dinosaurian endothermy. – In: THOMAS, R. D. K. & OLSON, E. C. [eds.] A cold look at the warm-blooded dinosaurs. – AAAS Selected Symposium **28**: 287-310.

JANENSCH, W. (1935-36): Die Schädel der Sauropoden *Brachiosaurus*, *Barosaurus* und *Dicraeosaurus* aus den Tendaguru-Schichten Deutsch-Ostafrikas. – Palaeontographica, Supplement **7**: 147-298.

– (1950): Die Skelettrekonstruktion von *Brachiosaurus brancai*. – Palaeontographica, Supplement **7**: 97-103.

LAPPARENT, A. F. & ZBYSZEWSKI, G. (1957): Les dinosauriens du Portugal. – Memórias dos Serviços geológicos de Portugal, (nova Sér.) **2**: 1-63.

MADSEN, J. H. (1974): A new theropod dinosaur from the Upper Jurassic of Utah. – Journal of Paleontology **48**: 27-31.

MARTIN, L. D. (1983): The origin and early radiation of birds: 291-338. – In: BRUSH, A. H. & CLARK, S. A. [eds.] Perspectives of Ornithology, New York (Cambridge University Press).

MAXWELL, W. D. & OSTROM, J. H. (1995): Taphonomy and paleobiological implications of *Tenontosaurus-Deinonychus* associations. – Journal of Vertebrate Paleontology **15**: 707-712.

MCINTOSH, J. S. (1990): Sauropoda: 345-401. – In: WEISHAMPEL, D. B., DODSON, P. & OSMÓLSKA, H. [eds.] The Dinosauria. – Berkeley (University of California Press).

MOLNAR, R. E., KURZANOV, S. M. & DONG, Z. (1990): Carnosauria: 169-209. – In: WEISHAMPEL, D. B., DODSON, P. & OSMÓLSKA, H. [eds.] The Dinosauria. – Berkeley (University of California Press).

OSTROM, J. H. (1969): Osteology of *Deinonychus antirrhopus*, an unusual theropod from the Lower Cretaceous of Montana. – Bulletin of the Peabody Museum of Natural History **30**: 1-165.

– (1970): Stratigraphy and paleontology of the Cloverly Formation (Lower Cretaceous) of the Bighorn Basin Area, Wyoming and Montana. – Bulletin of the Peabody Museum of Natural History **35**: 1-234.

– (1976): *Archaeopteryx* and the origin of birds. – Biological Journal of the Linnean Society 8: 91-182.

– (1978): The osteology of *Compsognathus longipes* WAGNER. – Zitteliana **4**: 73-118.

RAUHUT, O. W. M. & KRIWET, J. (1994): Teeth of a big theropod dinosaur from Porto das Barcas (Portugal). – Berliner geowissenschaftliche Abhandlungen **E 13**: 179-185.

RAUHUT, O. W. M. & ZINKE, J. (1995): A description of the Barremian dinosaur fauna from Uña with a comparison to that of Las Hoyas: 123-126. – In: II International Symposium on Lithographic Limestones, Lleida-Cuenca (Spain), Extended Abstracts, Madrid (Universidad Autónoma de Madrid).

SERENO, P. C. (1997): The origin and evolution of dinosaurs. – Annual Reviews of Earth and Planetary Sciences **25**: 435-489.

STROMER, E. (1934): Die Zähne des *Compsognathus* und Bemerkungen über das Gebiss der Theropoda. – Centralblatt für Mineralogie, Geologie und Palaeontologie B, **1934**: 74-85.

SUES, H.-D. & NORMAN, D. B. (1990): Hypsilophodontidae, *Tenontosaurus*, Dryosauridae: 498-509. – In: WEISHAMPEL, D. B., DODSON, P. & OSMÓLSKA, H. [eds.]: The Dinosauria; Berkeley (University of California Press).

THULBORN, R. A. (1973): Teeth of ornithischian dinosaurs from the Upper Jurassic of Portugal. – Memória dos Servicos Geológicos de Portugal (Nuova Seria) **22**: 89-134.

– (1992): Taxonomic characters of *Fabrosaurus australis*, an ornithischian dinosaur from the Lower Jurassic of southern Africa. – Geobios **25**: 283-292.

WAGNER, A. (1861): Neue Beiträge zur Kenntniss der urweltlichen Fauna des lithographischen Schiefers. V. *Compsognathus longipes* WAGN. – Abhandlungen der bayerischen Akademie der Wissenschaften **9**: 30-38.

WEIGERT, A. (1995): Isolierte Zähne von cf. *Archaeopteryx* sp. aus dem oberen Jura der Kohlengrube Guimarota (Portugal). – Neues Jahrbuch für Geologie und Paläontologie, Monatshefte **1995**: 562-576.

WEISHAMPEL, D. B. (1984): Evolution of jaw mechanisms in ornithopod dinosaurs. – Advancements in Anatomy, Embryology and Cell Biology **87**: 1-110.

– (1990): Dinosaurian distribution: 63-139. – In: WEISHAMPEL, D. B., DODSON, P. & OSMÓLSKA, H. [eds.] The Dinosauria; Berkeley (University of California Press).

WIECHMANN, M. F. & GLOY, U. (2000*): Pterosaurs and urvogels from the Guimarota mine. – In: MARTIN, T. & KREBS, B. [eds.] Guimarota – a Jurassic ecosystem: 83-86, München (Verlag Dr. F. Pfeil).

ZINKE, J. (1998): Small theropod teeth from the Upper Jurassic coal mine of Guimarota (Portugal). – Paläontologische Zeitschrift **72**: 179-189.

ZINKE, J. & RAUHUT, O. W. M. (1994): Small theropods (Dinosauria, Saurischia) from the Upper Jurassic and Lower Cretaceous of the Iberian peninsula. – Berliner geowissenschaftliche Abhandlungen **E 13**: 163-177.

Pterosaurs and urvogels from the Guimarota mine

MARC FILIP WIECHMANN & UWE GLOY

Today, birds and bats represent the actively flying vertebrates. However, birds are only known in higher numbers since the Early Cretaceous and bats first appear in the fossil record in the Early Tertiary. In the Late Jurassic, the pterosaurs were the dominant flying vertebrates. Although a relative of the urvogel *Archaeopteryx* is known from the sediments of the Guimarota mine, it certainly did not represent a serious competition for the pterosaurs. Pterosaurs are represented in the Guimarota mine with more than 300 isolated teeth and a few articulated and disarticulated skeletal remains. Archaeopterygiform birds, in contrast, are "only" known from 103 teeth (WEIGERT 1995).

Pterosauria

Pterosaurs are known in the fossil record since the Late Triassic (Norian; WILD 1978), and their last representatives died out at the end of the Cretaceous. They are the first actively flying animals among the vertebrates. Since HAMMER & HICKERSON (1993) reported pterosaurs from Antarctica, they are known to be present on all continents (LAWSON 1975, WILD 1978, WELLNHOFER 1991a,b, UNWIN et al. 1996, UNWIN & HEINRICH 1999). Active flying requires the transformation of the forelimbs into wings and a significant reduction of the weight of the skeleton. The latter was achieved by the development of hollow bones, a very light construction of the skull and the reduction of other skeletal elements (WELLNHOFER 1978).

The first primitive pterosaurs belong to a paraphyletic group commonly known as "Rhamphorhynchoidea" (UNWIN 1995), which died out in the middle Early Cretaceous (JI et al. 1999, SWISHER et al. 1999). The more modern pterosaurs – the monophyletic Pterodactyloidea – are known since the Late Jurassic (UNWIN 1995), and they dominated the airspace until the end of the Cretaceous.

Pterosaurs from the Guimarota mine

Apart from teeth, no elements of the skull are preserved. The preserved teeth exhibit the typical characters of pterosaur teeth (WELLNHOFER 1978). They are more or less strongly recurved, slightly flattened transversely, smooth and pointed towards their tip (Fig. 12.1). Some of the teeth exhibit an enamel cap, from which a band of enamel runs to the anterior and posterior side of the tooth, respectively. Together with the flexure of the teeth, this character indicates that these specimens represent a relative of the genus *Rhamphorhynchus*. In contrast, other teeth show crowns that are completely covered with enamel and very fine striations. They can be referred to the Pterodactyloidea (UNWIN, pers. com. 1999).

Only very fragmentary, though partially articulated remains of the postcranium are present. Apart from many fragments of the hollow long bones, which cannot be identified, a left pectoral girdle and several digits of the hand can be referred to the Pterosauria. The pectoral girdle (Fig. 12.2) is very similar in its dimensions to the scapulacoracoid of *Rhamphorhynchus muensteri* and thus indicates an animal with a wing span of approximately 1 m (WELLNHOFER 1975a, UNWIN, pers. com. 1999). A more precise determination of the systematic position of all of these specimens within pterosaurs is unfortunately not possible, due to their poor preservation. The only exception is the terminal phalanx of a flight finger (digit four) with a conspicuous flexure at its distal end. This character is typical for the Rhamphorhynchinae. In addition, a longitudinal furrow is preserved, which served as the attachment area for the wing membrane. Such a feature is only known from the genera *Rhamphorhynchus* and *Nesodactylus* within the Rhamphorhynchinae (WELLNHOFER 1975a, UNWIN, pers. com. 1999). No remains of pterodactyloids have been found in the postcranial material so far, but their presence is demonstrated by the presence of their teeth.

Fig. 12.1. Lateral view of a pterosaur tooth (Gui Pt 32) from the Guimarota mine; height of the tooth 7 mm.

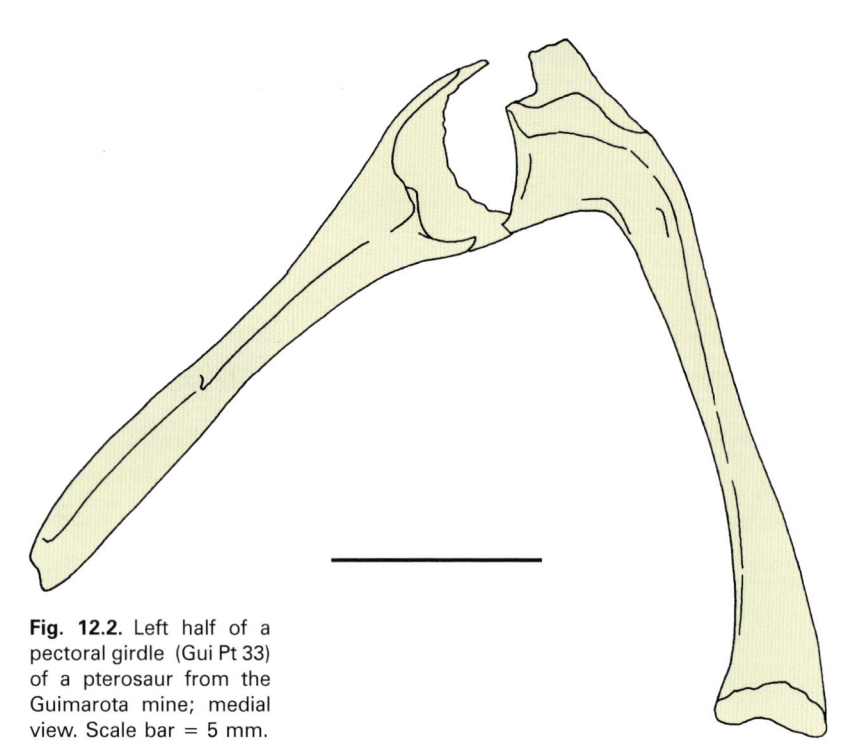

Fig. 12.2. Left half of a pectoral girdle (Gui Pt 33) of a pterosaur from the Guimarota mine; medial view. Scale bar = 5 mm.

Lifestyle of the pterosaurs

Pterosaur remains are mainly found in marine sediments, which gives an indication of the preferred habitat and food sources of these animals. Discoveries of *Rhamphorhynchus* (WELLNHOFER 1975b) and *Pterodactylus* (BROILI 1938) with preserved stomach- or throat-pouch-contents give testimony of the piscivorous (fish-eating) diets in many cases; this is also indicated by the dentitions. WELLNHOFER (1978) interpreted the sharply pointed, anteriorly directed teeth in *Ramphorhynchus* as an excellent tool for catching and securing fishes in the snout. In addition, the stocky skull and short neck formed an almost horizontal functional unit for capturing prey out of the water during flight.

The flora, fauna, sedimentology and coal petrology show that the Guimarota beds represent a coastal swamp with temporary marine influences, which was probably similar to Recent mangrove environments (GLOY 2000*). However, it is questionable whether the rhamphorhynchoids really lived in this environment. The long tail was not suited for a head-down hanging resting position from the branches of trees or large shrubs (WELLNHOFER 1978), although a life on the treetops might be possible. However, even if they did not live in the coastal swamp, the fish-eating pterosaurs probably hunted above the open water areas and lagoonal lakes.

Archaeopterygiformes (relatives of *Archaeopteryx*)

So far, *Archaeopteryx* has only been identified from the Solnhofen plattenkalks of the Late Jurassic of southern Germany with certainty. The skeletons of the world-famous urvogel have been described in detail in many publications (DEBEER 1954, HELLER 1959, WELLNHOFER 1974 and 1993). The teeth of *Archaeopteryx* were studied by HOWGATE (1984) and WELLNHOFER (1974 and 1993), who first figured finest structures of the teeth of *Archaeopteryx* with the help of a scanning electron microscope.

cf. *Archaeopteryx* in the Guimarota mine

WEIGERT (1995) described isolated teeth from the Guimarota mine, which are similar to the teeth of *Archaeopteryx lithographica* and *Archaeopteryx bavarica*. The laterally convex teeth have a well-developed sigmoidal shape and are significantly broader at their basis than at their tip (Fig. 12.3). A conspicuous constriction is present in the middle part of the crown and between crown and root. The cutting edges show a fine serration in their upper parts. This area is not visible in the specimens of *Archaeopteryx* from Solnhofen, since they are still embedded in the sediment, so that the character of the serration cannot be compared between the specimens from the two localities. The cross-section of the teeth is rounded. With an average height of 1.65 mm, the teeth from the Guimarota mine are considerably larger than those of *Archaeopteryx* from the Solnhofen plattenkalks (WEIGERT 1995). Because of the significant morphological similarity, the S-shaped teeth from the Guimarota mine were referred to the Archaeopterygiformes by WEIGERT (1995). Thus, these teeth represent the oldest occurrence of this group.

Lifestyle

Of course, nothing can be said about the possible flight capabilities of *Archaeopteryx* on the basis of the material from the Guimarota mine. However, the reconstruction of the habitat allows indirect conclusions about the lifestyle of the "urvogel". If Guimarota is regarded as a mangrove-like swamp, both a terrestrial, as well as an arboreal lifestyle is possible. Like modern ground-dwelling birds, *Archaeopteryx* might have hunted its prey (e.g. small amphibians, reptiles and mammals) on dry, solid areas with lush vegetation (WELLNHOFER 1993, CHIAPPE 1997). However, the branches of small groves of trees, which housed possible prey items, would also be a suitable habitat for a tree climber (BOCK 1985, YALDEN 1985, FEDUCCIA 1993). Its more or less well developed flight capabilities (CAPPLE et al. 1983, OSTROM 1985, PADIAN 1985) would have enabled *Archaeopteryx* to cross larger glades, hunt flying prey, or escape from predators. In a mainly ground-dwelling lifestyle, *Archaeopteryx* would have mainly been hunted by the numerous crocodiles and predatory dinosaurs (theropods) (RAUHUT 2000*).

Taphonomy

The adaptations for flight led to a weight saving construction of the skeleton, which, however, is disadvantageous for its fossilization potential. Hollow bones are more easily compressed and destroyed during diagenesis than those that have a thick outer compact layer. Therefore, the flying vertebrates are mainly represented by teeth, which are the hardest and most resistant elements of the organism. The continuous tooth replacement in these animals also played a role in the accumulation of the teeth. A further explanation might be that the slender teeth often broke during hunting.

Because of their pneumatised, hollow bones, the corpses of pterosaurs and *Archaeopteryx* floated on the water surface for a long time, and did not sink to the ground rapidly. The buoyancy of the urvogels was further enhanced by the air that was trapped in the feathers. During this time, the corpses were easy prey for crocodiles, fishes and other scavengers. If the corpses floated on the surface untouched for a longer time, first the peripheral skeletal elements, such as the skull, neck, limbs, and tail, were separated from the rest of the skeleton and buried in isolation.

Acknowledgements

The authors would like to thank Dr. D. UNWIN, Berlin, for critical reading the manuscript and for his help in identifying the pterosaur remains.

References

BOCK, W. (1985): The arboreal theory of the origin of birds. – In: HECHT, M. K., OSTROM. J. H., VIOHL, G. & WELLNHOFER, P. [eds.] The beginning of birds. Proceedings of the International Archaeopteryx Conference Eichstätt: 199-207, Eichstätt (Freunde des Jura-Museums Eichstätt).
BROILI, F. (1938): Beobachtungen an *Pterodactylus*. – Sitzungsberichte der Bayerischen Akademie für Wissenschaften, mathematisch-naturwissenschaftliche Klasse **16**: 139-154.

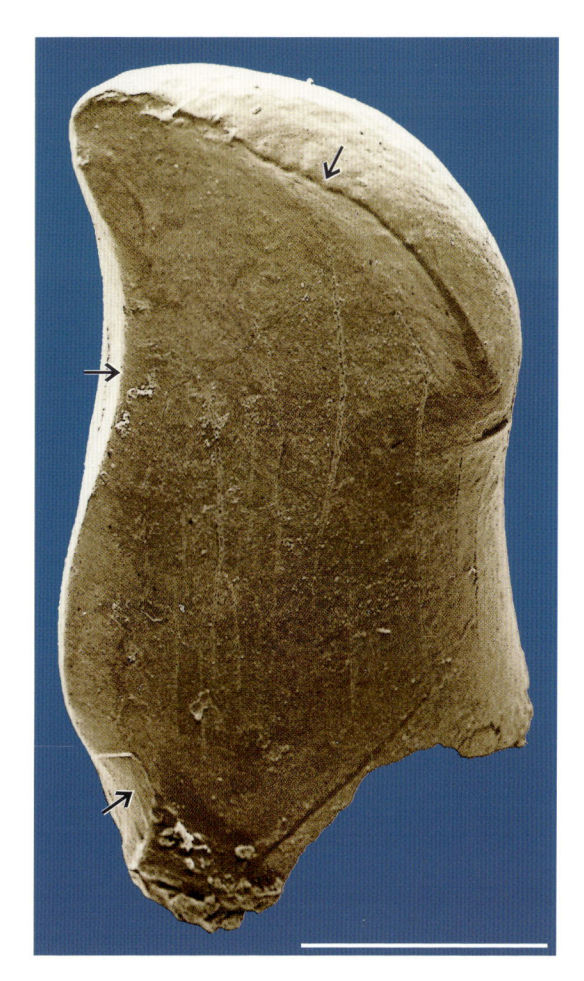

Fig. 12.3. Tooth of the right upper jaw of cf. *Archaeopteryx* (Gui Arch 10). The two arrows on the left indicate the constrictions, the arrow at the top indicates the serrations on the cutting edge. Scale bar = 0.5 mm.

CAPLE, G. R., BALDA, R. T. & WILLIS, W. R. (1983): The physics of leaping animals and the evolution of preflight. – American Naturalist **121**: 455-467.
CHIAPPE, L. M. (1997): Climbing *Archaeopteryx*? A Response to YALDEN. – Archaeopteryx **15**: 109-112.
DEBEER, G. R. (1954): *Archaeopteryx lithographica*: a study based on the British Museum specimen. – 68 pp., London (British Museum [Natural History]).
FEDUCCIA, A (1993): Evidence from claw geometry indicating arboreal habits of *Archaeopteryx*. – Science **259**: 790-793.
GLOY, U. (2000*): Taphonomy of the Guimarota beds. – In: MARTIN, T. & KREBS, B. [eds.] Guimarota – a Jurassic ecosystem: 129-135, München (Verlag Dr. F. Pfeil).
HAMMER, W. R. & HICKERSON, W. J. (1993): A new Jurassic dinosaur fauna from Antarctica. – Journal of Vertebrate Paleontology **13**, 3 (Suppl.): 40A.
HELLER, F. (1959): Ein dritter *Archaeopteryx*-Fund aus den Solnhofer Plattenkalken von Langenaltheim/Mfr. – Erlanger Geologische Abhandlungen **31**: 1-25.
HOWGATE, M. E. (1984): The teeth of *Archaeopteryx*, and reinterpretation of the Eichstätt specimen. – Zoological Journal of Linnean Society **82**: 159-175.
JI, S.-A., JI, Q. & PADIAN, K. (1999): Biostratigraphy of new pterosaurs from China. – Nature **398**: 573-574.

Lawson, D. A. (1975): Pterosaur from the Latest Cretaceous of West Texas. Discovery of the largest flying creature. – Science **187**: 947-948.

Ostrom, J. H. (1985): The meaning of *Archaeopteryx*. – In: Hecht, M. K., Ostrom, J. H., Viohl, G. & Wellnhofer, P. [eds.] The beginning of birds. – Proceedings of the International Archaeopteryx Conference Eichstätt: 161-176, Eichstätt (Freunde des Jura-Museums Eichstätt).

Padian, K. (1985): The origins and aerodynamics of flight in extinct vertebrates. – Palaeontology **28**: 413-433.

Rauhut, O. W. M. (2000*): The dinosaur fauna from the Guimarota mine. – In: Martin, T. & Krebs, B. [eds.] Guimarota – a Jurassic ecosystem: 75-82, München (Verlag Dr. F. Pfeil).

Swisher III, C. C., Wang, Y.-Q., Wang, X.-L., Xu, X. & Wang, Y. (1999): Cretaceous age for the feathered dinosaurs of Liaoning, China. – Nature **400**: 58-61.

Unwin, D. M. (1995): Preliminary results of a phylogenetic analysis of the Pterosauria (Diapsida: Archosauria). – In: Sun, A. & Wang, Y. [eds.]: Sixth Symposium on Mesozoic Terrestrial Ecosystems and Biota, Short Papers: 69-72, Beijing (China Ocean Press).

Unwin, D. M., Manabe M., Shimizu, K. & Hasegawa, Y. (1996): First record of pterosaurs from the Early Cretaceous Tetori Group: a wing-phalange from the Amagodani Formation in Shokawa, Gifu Prefecture, Japan. – Bulletin of National Science Museum, Tokyo, Series C **22**: 37-46.

Unwin, D. M. & Heinrich, W.-D. (1999): On a pterosaur jaw from the Upper Jurassic of Tendaguru (Tanzania). – Mitteilungen aus dem Museum für Naturkunde Berlin, Geowissenschaftliche Reihe **2**: 121-134.

Weigert, A. (1995): Isolierte Zähne von cf. *Archaeopteryx* sp. aus dem Oberen Jura der Kohlengrube Guimarota. – Neues Jahrbuch für Geologie und Paläontologie, Monatshefte **9**: 562-576.

Wellnhofer, P. (1974): Das fünfte Exemplar von *Archaeopteryx*. – Palaeontographica A **147**: 169-216.

– (1975a): Die Rhamphorhynchoidea (Pterosauria) der Oberjura-Plattenkalke Süddeutschlands. I. Allgemeine Skelettmorphologie. – Palaeontographica A **148**: 1-30.

– (1975b): Die Rhamphorhynchoidea (Pterosauria) der Oberjura-Plattenkalke Süddeutschlands. III. Palökologie und Stammesgeschichte. – Palaeontographica A **149**: 1-30.

– (1978): Pterosauria. – In: Wellnhofer, P. [ed.] Handbuch der Paläoherpetologie **19**: 1-82, Stuttgart (Gustav Fischer Verlag).

– (1991a): The illustrated encyclopedia of pterosaurs. – 192 pp., London (Salamander Books).

– (1991b): Weitere Pterosaurierfunde aus der Santana-Formation (Apt) der Chapada do Araripe, Brasilien. – Palaeontographica A **215**: 43-101.

– (1993): Das siebte Exemplar von *Archaeopteryx* aus den Solnhofer Schichten. – Archaeopteryx **11**: 1-48.

Wild, R. (1978): Die Flugsaurier (Reptilia, Pterosauria) aus der Oberen Trias von Cene bei Bergamo, Italien. – Bollettino della Societá Paleontologica Italiana **17**: 176-259.

Yalden, D. (1985): Forelimb function in *Archaeopteryx*. – In: Hecht, M. K., Ostrom, J. H., Viohl, G. & Wellnhofer, P. [eds.] The beginning of birds. – Proceedings of the International Archaeopteryx Conference Eichstätt: 91-98, Eichstätt (Freunde des Jura-Museums Eichstätt).

Eggshells from the Guimarota mine

Rolf Kohring

Our knowledge about the reproductive strategies of extinct groups of vertebrates is unfortunately poor, since most of the important information does not become fossilized. Embryos are known from a few, exceptionally preserved vertebrate fossils, like the Liassic ichthyosaurs from Holzmaden or some Early Tertiary horses. Complete eggs and nests are the only available evidence for other groups. Such discoveries are spectacular, but unfortunately very rare. Isolated eggshells, which are preserved in suitable sediments, however, are relatively common, although they have long been ignored. Since the evolutionary development of a hard eggshell was linked to the invasion of the dry land, such eggshells are almost always found in non-marine, usually limnic or fluviatile sediments. Here, they occur together with other calcareous microfossils, such as ostracodes, charophytes, and snails. In many well known vertebrate localities of the Cenozoic, remains of eggshells are quite common, including the Geiseltal (Middle Eocene) (KOHRING & HIRSCH 1996), Sieblos in the Rhön (Lower Oligocene) (MARTINI 1988) and the Nördlinger Ries (Miocene) (KOHRING & SACHS 1997). Amongst Mesozoic localities that have yielded eggshells, the Late Jurassic Morrison Formation (USA) (HIRSCH 1994), the Early Cretaceous of Spain (Galve, Uña) (KOHRING 1991) and the Late Cretaceous of southern France (DUGHI & SIRUGUE 1976, ERBEN et al. 1979) might be mentioned. The Guimarota mine has also yielded some remains of eggshells.

Today, eggs and eggshells can often be identified on the basis of their outer features, such as color, or possibly surface structures. If the egg is complete, the size and shape can also be used for identification. So what can be said about the fragments of fossil eggshells, which often seem to be insignificant at first sight? The shape and size of the eggs can almost never be determined, the color is gone, and color patterns are very rarely preserved and are only visible under ultraviolet light. The thickness of the shell is also not very useful, if it is not at least roughly known which animals laid the eggs. Thus, the thickness of the eggshells in modern birds can vary between 28 μm (hummingbirds) and almost 3 mm (African ostrich). Interestingly, however, there is one important character that can be studied even in tiny fragments, which is the shell structure itself (Fig. 13.1). If a calcareous eggshell of a modern bird is studied in cross-section, under the scanning electronic microscope, or in thin section, vertically oriented, cylindrical units can be recognized, which are attached to a layer of organic fibers, the so-called Membrana testacea. These calcareous units can vary significantly in their dimensions, the arrangements of single structured zones and in their mineralogy. Eggshells of crocodiles, dinosaurs and birds are made up of calcite and exhibit special characters for each group. Some lizards, e.g. the geckos, also use calcite to build their eggshells, usually in homogeneous forms or in

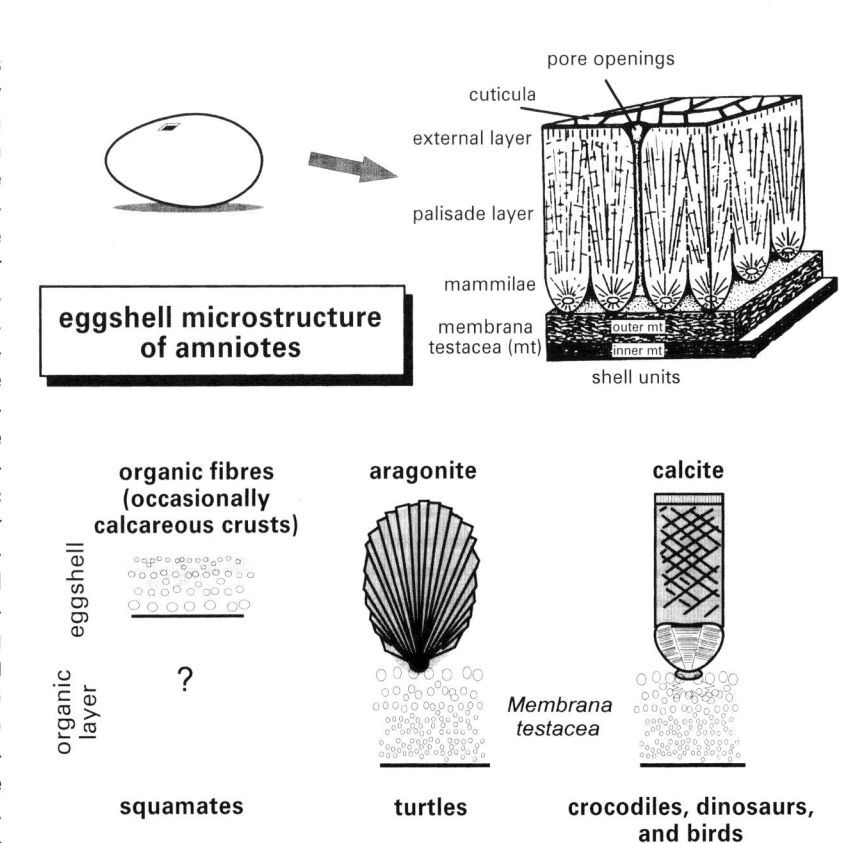

Fig. 13.1. Structure of the eggshells of amniotes. The eggshells of turtles and archosaurs (= crocodiles, dinosaurs, and birds) consist of cylindrical units. Only turtles form these units from aragonite.

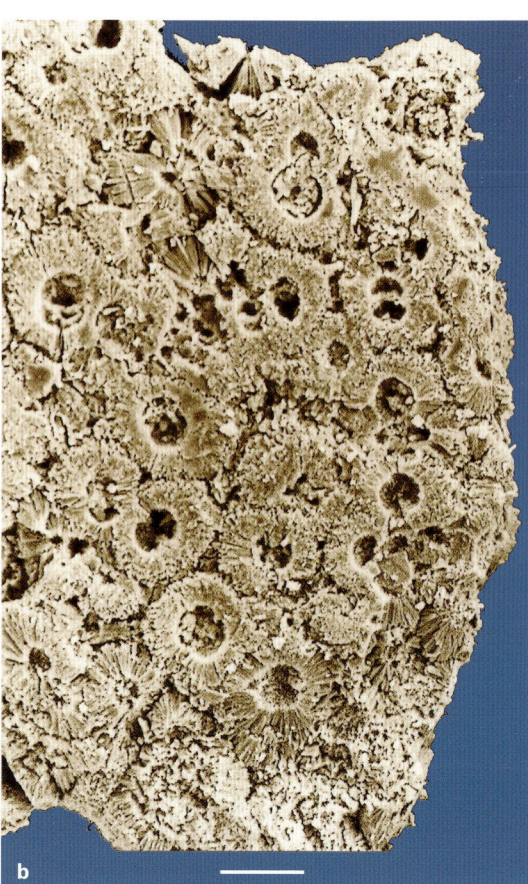

Fig. 13.2. Fossil turtle eggshell *(Testudoolithus)* from the Late Jurassic of the Guimarota mine (**a**: radial view; **b**: internal view). Scale bar = 100 μm.

radial units. Only turtles exhibit a slight diversion from this pattern; they do build their eggshells in the form of cylindrical units, but, as the only group of amniotes, they use the more instable variety of lime, the aragonite. Since most of the visible characters are genetically controlled (a strong ecological control seems only to be present in the soft-shelled eggs of snakes and many lizards), fossil remains can, in the light of our knowledge of Recent eggshells, be referred to different groups of animals with some certainty.

During his studies on the ostracodes of the Guimarota mine, HELMDACH (1971) found some eggshell fragments in the layer KM 11, at the basis of the profile he worked on. He picked these eggshell fragments out and also mentioned them in his paper in passing. A first description of this material was not published until much later (KOHRING 1990). The superbly preserved eggshells are approximately 150 μm thick (Fig. 13.2). Starting from the basal primary spherites, i.e. the central cores with a diameter of c. 25 μm, the shell is formed by aragonitic, needle-like crystals. The single units of the shell are very slender, their diameter is only 50 μm. Thus, the ratio between the width and the height of the eggshell units is approximately 1:3. The outer surface of the shell is characterized by the wavy elevation of the single units. In thin section, tightly spaced concentric growth lines are visible. Under polarized light, weak pleochroism is visible in the units of the shell; this is typical for eggshells that have not been subject to great diagenetic stresses (HIRSCH 1983). All of the characters, but especially the aragonitic shell units, demonstrate conclusively that these eggshells belong to turtles. Based on the conspicuous ratio between width and height of the shell units, the eggs can be identified as being laid by terrestrial turtles. For comparison, eggshells of Recent tortoises (*Testudo horsfieldii* GRAY 1844, the four-toed tortoise, and *Testudo pardalis* BELL 1828, the leopard tortoise) are illustrated here next to the shells from the Guimarota mine. In these taxa, the ratio between width and height of the units varies between 1:2 and 1:2.7. However, no skeleton remains or parts of the theca of truly terrestrial turtles have been described from Guimarota so far.

Later, other eggshell remains have been found in the material from Guimarota, which can also be referred to turtles. They are similar to the shells described above in their mineralogical structure, but differ in their dimensions. Although they have almost the same shell thickness, the ratio between width and height of the single units is 1:1, which is typical for the eggshells of many aquatic turtles (KOHRING 1998).

The find of turtle eggshells with preserved aragonite from the Late Jurassic is quite surprising, since aragonite is much more likely to be dissolved than calcite, and is transformed into the latter during recrystallisation (by which the original structures are often lost). HIRSCH (1983) distinguished three different stades of preservation in fossil turtle eggshells: unaltered specimens, partially altered specimens (mainly aragonite, but with some substitutions by calcite in the pore channels and in the basal areas of the shell units) and finally the completely altered shells (shell made up of large crystals of calcite). TUREKIAN & ARMSTRONG (1961) already demonstrated that fossil mollusk shells that are preserved in the original aragonite are not necessarily completely unaltered in respect to their microstructure and mineralogy. They found significant contamination with rare elements and speculated that a secondary mineralization happens in spaces becoming available by the decay of the organic matrix (up to 80 %), which then helps in the preservation of the aragonite. WEINER & LOWENSTAM (1981) showed that the proteins of the organic matrix have been transformed into short chains of peptides in well preserved Cretaceous bivalve shells. Accordingly, CLARK (1991) assumed that at some time during the breakdown of the organic matrix, the remaining matrix absorbs traces of metals such as iron and copper, and thus preserves the aragonite crystal unaltered. Because of the special preservational conditions in the Guimarota mine, the turtle eggshells obviously underwent similar preservational processes. They are thus even better preserved than remains from many Tertiary localities.

Today, fossil eggshells are classified in a parataxonomy, as it is the case for trace fossils. Thus, "generic" and "specific" names are given on the basis of the observed characters of the specimens, and not on the basis of the usually unknown producers of the eggs. The parataxonomic name *Testudoolithus* was introduced by HIRSCH (1996) for fossil tortoise eggshells. This type of eggshell is common, especially in Cenozoic sediments. The fossil eggshells from the Guimarota mine can also be classified in a parataxonomy.

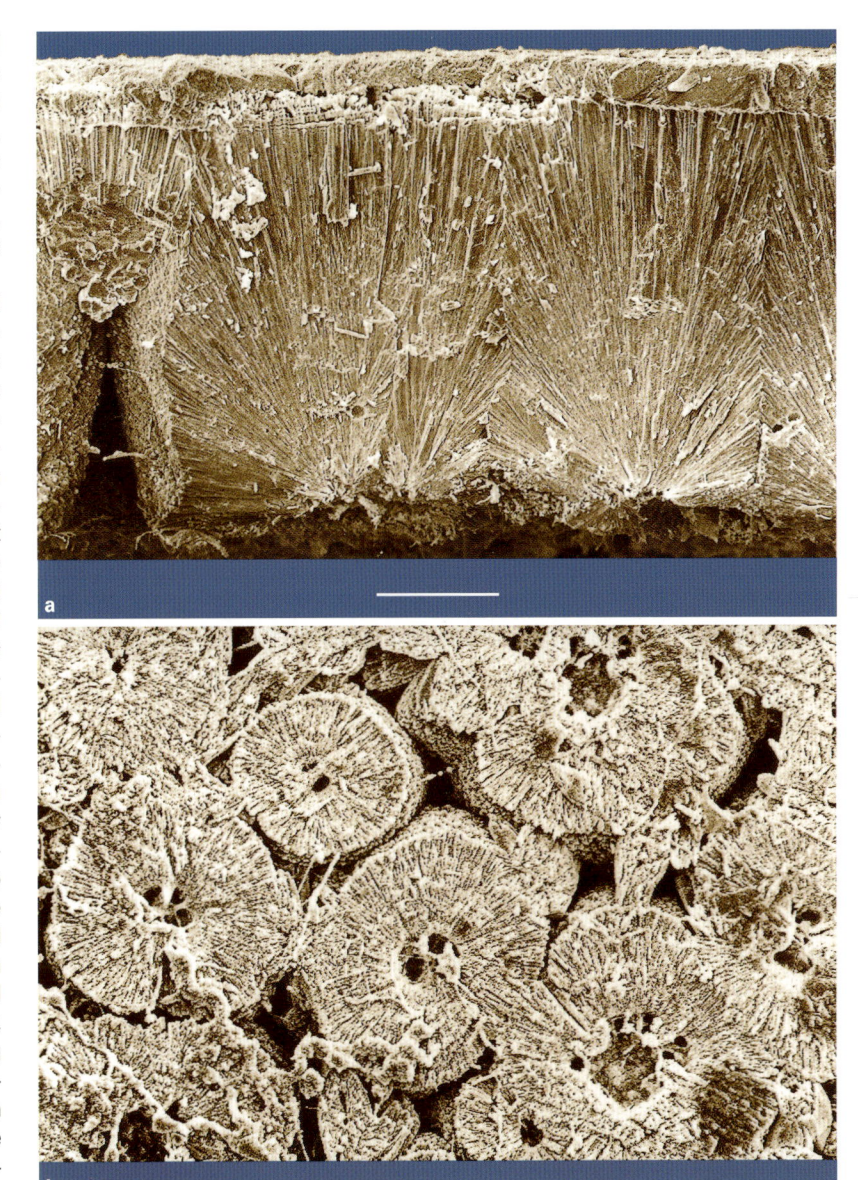

The eggshells of the terrestrial turtles can be referred to the genus *Testudoolithus* (KOHRING 1999) and thus represent the oldest known fossil eggshells of tortoises that can be identified with certainty, while the eggshells of the aquatic animals are named *Chelonoolithus* (KOHRING 1998). No eggshells of other groups, such as crocodiles or dinosaurs, have been found in the Guimarota mine so far, although they are to be expected. The eggshells found so far are certainly allochthonous, i.e. they have been transported over some distance, since turtles always lay their eggs on dry land. Because of their specific weight and resulting differences in buoyancy, the calcitic eggshells

Fig. 13.3. Modern turtle eggshell. Radial view of *Testudo pardalis* (**a**) and internal view of an eggshell of *Testudo horsfieldii* (**b**). Scale bar = 100 μm.

of other groups might thus have been buried in other areas within the depositional environment of Guimarota.

References

BELL, T. (1828): On three new species of land tortoises. – Zoological Journal **3**: 420-421.

CLARK, G. R. (1991): Physical evidence for organic matrix degradation in fossil mytilid (Mollusca: Bivalvia) shells. – In: KOBAYASHI, I., MUTVEI, H. & SAHNI, A. [eds.] Structure, formation and evolution of fossil hard tissue: 73-79, Tokyo (Tokyo University Press).

DUGHI, R. & SIRUGUE, F. (1976): L'extinction des dinosaures à la lumière des gisements d'oeufs du Crétacé terminal du Sud de la France, principalement dans le bassin d'Aix-en-Provence. – Paléobiologie continentale **7**: 1-39.

ERBEN, H. K., HOEFS, J. & WEDEPOHL, K. H. (1979): Paleobiological and isotopic studies of eggshells from a declining dinosaur species. – Paleobiology **5**: 380-414.

GRAY, E. (1844): Catalogue of the tortoises, crocodyles, and amphisbaenians in the collections of the British Museum **7**: VIII, 80 pp., London (British Museum [Natural History]).

HELMDACH, F.-F. (1971): Stratigraphy and ostracode-fauna from the Coalmine Guimarota (Upper Jurassic). – Memórias dos Serviços Geológicos de Portugal, N.S. **17**: 43-88.

Hirsch, K. F. (1983): Contemporary and fossil chelonian eggshells. – Copeia **1983**: 382-397.

– (1994): The fossil record of vertebrate eggs. – In: DONOVAN, S. K. [ed.] The paleobiology of trace fossils: 269-294, Chichester, England (Wiley and Sons).

– (1996): Parataxonomic classification of fossil chelonian and gecko eggs. – Journal of Vertebrate Paleontology **16**: 752-762.

KOHRING, R. (1990): Upper Jurassic chelonian eggshell fragments from the Guimarota coalmine (Central Portugal). – Journal of Vertebrate Paleontology **10**: 128-130.

– (1991): Lizard egg shells from the Lower Cretaceous of Cuenca Province, Spain. – Palaeontology **34**: 237-240.

– (1998): Neue Schildkröten-Eischalen aus dem Oberjura der Grube Guimarota (Portugal). – Berliner geowissenschaftliche Abhandlungen E **28**: 113-117.

– (1999): Strukturen, Biostratinomie und systematische und phylogenetische Relevanz von Eischalen amnioter Wirbeltiere. – Courier Forschungsinstitut Senckenberg **210**: 1-311.

KOHRING, R. & HIRSCH, K. F. (1996): Crocodilian and avian eggs and eggshells from the Eocene of the Geiseltal. – Journal of Vertebrate Paleontology **16**: 67-80.

KOHRING, R. & SACHS, O. (1997): Erhaltungsbedingungen und Diagenese fossiler Vogeleischalen aus dem Nördlinger Ries (Miozän, MN6). – Archaeopteryx **15**: 73-96.

MARTINI, E. (1988): Ein Schildkröten-Ei aus dem Unter-Oligozän von Sieblos a.d. Wasserkuppe/Rhön. – Beiträge zur Naturkunde von Osthessen **24**: 169-173.

TUREKIAN, K. K. & ARMSTRONG, R. L. (1961): Chemical and mineralogical composition of fossil molluscan shells from the Fox Hill Formation, South Dakota. – Bulletin of the Geological Society of America **72**: 1817-1828.

WEINER, S. & LOWENSTAM, H. (1981): Well-preserved fossil mollusc shells: characterization of mild diagenetic processes. – In: HARE, P. E. [ed.] Biogeochemistry of amino acids: 95-114, New York (Wiley).

The docodont *Haldanodon* from the Guimarota mine

THOMAS MARTIN & MANUELA NOWOTNY

Four orders of Mesozoic mammals are present in the Guimarota mine: the Docodonta, Multituberculata, Dryolestida, and the Zatheria. Whereas the Dryolestida and Zatheria represent relatives of the modern mammals, the placental mammals and the marsupials, the Docodonta and Multituberculata are extinct groups. The geologically oldest docodonts are known from the Middle Jurassic (middle to late Bathonian) of the Isle of Skye (Scotland) (WALDMAN & SAVAGE 1972) and from Kirtlington (late Bathonian) in southern England (KERMACK et al. 1987). KERMACK et al. (1987) described a lower jaw fragment with one premolar and two molars from Kirtlington as a new genus and species, *Simpsonodon splendens*, which is very similar to the docodont from Guimarota; furthermore, several isolated molars from this locality are also present. The next youngest and at the same time best recorded docodonts are those from the Late Jurassic (Kimmeridgian) of the Guimarota mine (KRUSAT 1980, LILLEGRAVEN & KRUSAT 1991). The name bearing genus of this group, *Docodon*, is known from the slightly younger Morrison Formation of Wyoming with possibly five species; however, these taxa are only known from a few jaw fragments.

Haldanodon exspectatus

The Docodonta are known from Guimarota with only one genus and species, *Haldanodon exspectatus* KÜHNE & KRUSAT 1972. As is the case with the other mammals from Guimarota, this taxon is mainly known from teeth and jaws. However, apart from the approximately 200 jaw remains, ten more or less complete skulls of this mammal have also been found. An especially lucky find was the discovery of a partially preserved skeleton (Fig. 14.1), which is exceptionally for Mesozoic mammals. Although teeth are very important for the determination of the systematic position of a fossil mammal within the Mammalia, only the skull (Fig. 14.2) and especially the skeleton can provide us with information on the appearance and the lifestyle of the animal.

As is the case with most Mesozoic mammals, *Haldanodon* was a small animal, which approximately reached the size of a mole. Looking at the unfortunately incomplete and disarticulated skeleton, the unusually strong and massive limb bones are especially striking (Fig. 14.1). Especially the upper leg bone (femur) and the upper arm bone (humerus) are stocky and sturdy, with massive, strongly broadened articular ends. This is not the skeleton of an agile and nimble climber like *Henkelotherium*, but of a stocky ground dweller. In small vertebrates, such strong modifications always indicate a highly specialized lifestyle. A very vivid example for this is the mole, which is greatly adapted for a burrowing lifestyle, and in which the muscle attachment areas in the upper arm bone are strongly broadened. On the other hand, mice, which, as generalized animals, are able to dig, but also jump and often climb, do not exhibit any special adaptations in their skeleton.

A comparison of the preserved postcranial bones of *Haldanodon* with the corresponding skeletal elements of small, ground-dwelling, recent animals resulted in the recognition of great similarities with the skeleton of the desmane *(Desmana)*. Desmanes are semiaquatic representatives of the moles (Talpidae), and are known from one genus and species from northern Spain *(Galemys pyrenaicus)* and the Volga area *(Desmana moschata)*, respectively. They dig their burrows into the banks of streams and rivers and hunt small invertebrates such as snails, insects and their larvae, in the vicinity of the shores. The forelimbs are used for digging, whereas the hindlimbs with their elongated feet and the laterally flattened tail are used for swimming. Unfortunately, the feet and tail of *Haldanodon* are not preserved, so that nothing can be said about possible swimming adaptations in these parts of the body. The adaptations for digging, on the other hand, are clearly visible in the upper arm bone of *Haldanodon*, with its flattened shaft, the enormous crest for the attachment of the deltopectoralis musculature and the broad articular ends. However, the humerus is not as specialized as that of the mole, which is a highly specialized burrower and with

Fig. 14.1a. Incompletely preserved skeleton of *Haldanodon exspectatus*, including the skull and parts of the fore- and hindlimbs (Gui Mam 30/79). The upper arm bone (humerus) and upper leg bone (femur) are both massive and stocky. The broadened articular ends of the humerus served for the attachment of powerful muscles, which were used by *Haldanodon* for burrowing.

Fig. 14.1b. Drawing of the skeleton figured in 14.1a, labeling the skeletal elements. Fe – Femur (upper leg bone), Hu – Humerus (upper arm bone), Ju – Jugal, M^5 – fifth upper molar, Md – Mandible (lower jaw), Mx – Maxilla (upper jaw), Pmx – Premaxilla, Ra – Radius, Sc – Scapula (shoulder blade), Sq – Squamosum, Ti – Tibia (shin bone), Ul – Ulna. Drawing: P. BERNDT.

which *Haldanodon* has been compared previously (KRUSAT 1991). In the mole, the humerus is almost as broad as long, which is in strong contrast to the morphology of this bone in the docodont. A semiaquatic animal with a certain adaptation for burrowing fits much better into a coaly swamp environment such as Guimarota than a strict burrower such as the mole, which prefers dry, well drained substrates.

LILLEGRAVEN & KRUSAT (1991) analyzed the anatomy of the skull remains of *Haldanodon* in detail and presented a reconstruction of the skull. The dermal bones of the skull roof of *Haldanodon* are unusually massive and strongly mineralized (Fig. 14.2). The paired nasals form large, thick plates of bone, which covered almost the entire anterior half of the skull. The dermal bones of the skull exhibit a well-developed relief. This is an indication that *Haldanodon* probably had a keratinous shield on its forehead, which protected the skull during burrowing (Fig. 14.3). A very similar protective adaptation is found in a relative of the hedgehogs, *Pholidocerus*, from the Middle Eocene oil shales of Messel, near Darmstadt (KOENIGSWALD & STORCH 1983). In lateral view, the wedge-shape of the skull of *Haldanodon* becomes obvious. The posteriorly directed occipital region of the skull is especially high and broad and exhibits a well developed bony crest. This large area on the occipital face served as an attachment area for the powerful neck musculature, which enabled *Haldanodon* to powerfully move its head up and down. Similar to the Recent African golden moles *Chrysochloris* and *Amblysomus*, *Haldanodon* probably used its skull to move the earth, which had been loosened with the forelimbs, behind the body (KRUSAT 1991). The dentition of *Haldanodon* is in general accordance with this interpretation of its lifestyle: in many jaws from the Guimarota mine, the molars are worn down to tiny stumps (Fig. 14.5). This phenomenon is typical for animals that search for their food underground, or close to river banks, and thus regularly feed on prey animals that are contaminated by sand and dirt (worms, insect larvae). The sand grains, which consist of quartz, are considerably harder than the calcium phosphate of the enamel and thus act as an aggressive abrasive on the teeth.

The position of docodonts within Mammalia

As mentioned above, teeth are the most important elements for the identification of a fossil mammal and to determine its systematic position. The reason for this is that teeth are very complex structures; they exhibit a high number of characters, which furthermore do not show much variation within one species. Another aspect is the fact that enamel is the hardest substance in a vertebrate body, so that teeth have a high potential of becoming fossilized. Looking at the dentition of *Haldanodon* (Figs. 14.4 to 14.8), it is immediately obvious that it is heterodont; i.e., different categories of teeth can be distinguished, including incisors, canines, premolars, and molars. Thus, *Haldanodon* exhibits on of the important criteria for a referral to the Mammalia, since this group is characterized by a considerable differentiation of the dentition, in contrast to reptiles. Looking at the molars of *Haldanodon* in detail (Fig. 14.6 to 14.8), their complex structure becomes obvious, but they lack the basal tricuspid pattern of the teeth of the Theria. Thus, this taxon obviously represents a lineage that developed a complex pattern of cusps and cutting edges in its molars independently from the lineage leading to the modern mammals. The upper molars seem to be especially characteristic. They are strongly broadened transversely, a character that led to the name Docodonta ("beam-tooth") for the whole group. This broadening of the teeth led to an increase in the chewing area in the teeth of docodonts, which enabled the animals to effectively crush their food. However, the docodonts could not crush

Fig. 14.2. Skull of *Haldanodon exspectatus,* left in dorsal view and right in palatal view (Gui Mam 41/75). The snout points upwards. Although the skull is flattened by the pressure of the overlying sediment, its wedge-like shape is still visible. The powerful articular joints of the occiput are well visible in the palatal view. The nasals and frontals have a roughened surface, indicating that they were probably covered by keratinous shields (dorsal view). Length 3.8 cm.

Fig. 14.3. Possible life reconstruction of *Haldanodon exspectatus*. The animal was probably a semiaquatic burrower, which was similar to the Recent desmanes *(Desmana* and *Galemys)*, a group that is related to the moles (Talpidae). The keratinous shields on the frontal area are reconstructed on the basis of roughened bone surfaces on the nasals. Adaptations to burrowing habits are the dense, short fur, the missing auricle, and the small eyes. Unfortunately, nothing is preserved of the skeleton of the hands, feet and tail, so that the reconstruction of these areas is hypothetical, as is that of the soft parts. Drawing: E. GRÖNING.

Fig. 14.4. Left lower jaw of *Haldanodon exspectatus* in lateral view (Gui Mam 81/79) with the sharp canine, three premolars and five molars. The well-developed coronoid process served as an attachment area for powerful jaw muscles. Length 2.8 cm

their food in the same way as rodents or large herbivores (ungulates), since their jaws did only allow very restricted transversal or longitudinal movements.

A biometrical and morphological analysis of the 152 lower jaw dentitions and 67 upper jaw dentitions (NOWOTNY in prep.) of *Haldanodon exspectatus* confirmed the presence of only one species of docodonts in the Guimarota mine; no sexual dimorphism could be recognized, in contrast to earlier assumptions (KRUSAT 1980). On the basis of the new material, the complete tooth formula, which was hitherto unknown, can now be reconstructed: the docodont had 6 incisors, 1 canine, 3 premolars, and 5 molars in the upper jaw, and 4 incisors, 1 canine, 3 premolars, and 5-6 molars in the lower jaw. The high number of incisors and molars is a primitive character, while the low number of premolars can be regarded as a specialization.

Fig. 14.5. Left lower jaw of *Haldanodon exspectatus* in lingual view (Gui Mam 35/75) with four incisors, the large canine, three premolars, and five strongly abraded molars. The groove for the rudimentary MECKEL's cartilage is well visible; it runs along the lower margin of the jaw towards the symphysis. Length 3 cm.

Seventeen jaws of juveniles with partially preserved milk dentitions are present among the 219 jaws of *Haldanodon exspectatus* examined. *Haldanodon* already had a diphyodont tooth replacement, i.e. the incisors, the canine, and the premolars were replaced once in a lifetime, as it is the case in modern mammals. Thus, with the exception of the molars, which are not replaced, the teeth of the permanent dentition represent the second tooth generation. The replacement of the premolars proceeded from front to back, with the replacement of the third premolar being slightly

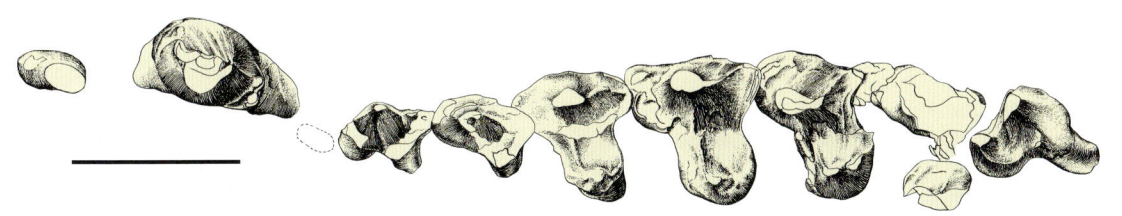

Fig. 14.6. Occlusal view of an upper jaw dentition of *Haldanodon exspectatus* (VJ 1008-155), exhibiting (from left to right) the large canine, two premolars and five molars; the fourth molar is fragmentary preserved. The striking broadening of the molars, transversely to the jaws, led to the name Docodonta (="beam-tooth"). Scale bar = 2 mm. From KRUSAT (1980).

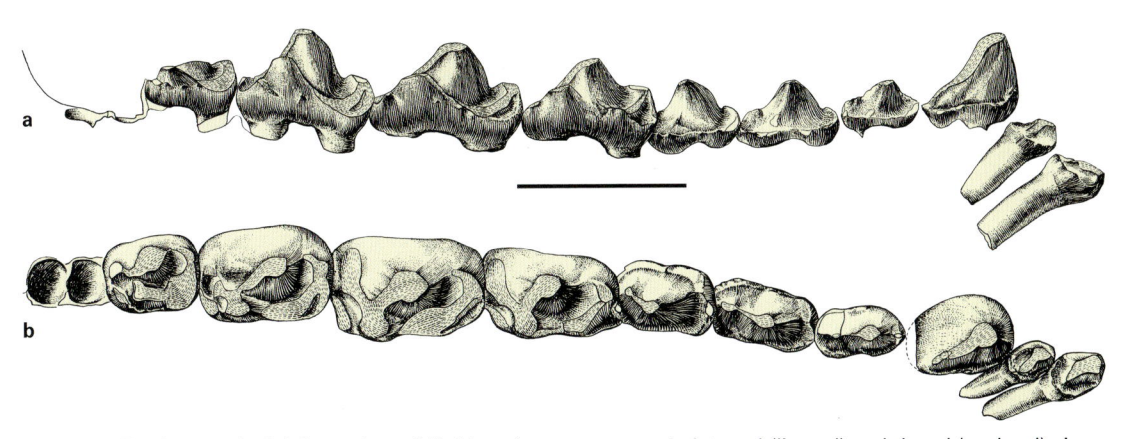

Fig. 14.7. Tooth row of a left lower jaw of *Haldanodon exspectatus* in internal (lingual) and dorsal (occlusal) views (VJ 1001-155, holotype). From right to left, two incisors, the canine, three premolars, and four molars are visible; the fifth molar has been lost. The docodonts developed a complex molar morphology independently from the eupantotheres. Especially the molars are strongly worn. Scale bar = 2 mm. From KRUSAT (1980).

Fig. 14.8. Scanning electron-microscopic photograph of the right lower dentition of *Haldanodon exspectatus* with P_1-P_3 and M_1-M_4. Behind M_4 the alveolus of M_5 is visible. (occlusal-lingual view) (Gui Mam 95/75). The molars exhibit a very complex pattern of cusps and cutting edges, while the structure of the premolars is rather simple. Scale bar = 2 mm.

delayed. The replacement of the premolars proceeds from front to back, with the replacement of the third premolar being slightly delayed: in Gui Mam 6/74 (lower dentition) M_{1-5} are fully erupted and dP_3 as the last milktooth is still in place; in Gui Mam 12/74 (upper dentition) at least M^{1-3} (jaw is broken behind M^3) are present when dP^2 and dP^3 are still in place. The canine is replaced after the P_2 in the lower jaw, and before the P^2 in the upper jaw. In both upper and lower jaws, M1 is fully erupted when the milk dentition is still complete. The milkteeth clearly differ in size and morphology from their permanent successors. In the lower jaw, dP_1 and dP_2 are premolariform and dP_3 is semimolariform; in the upper jaw, P^1 and P^2 are very small and peglike, while dP^3 is large and molariform. Since the milkteeth are replaced only once and a replacement of molariforms does not occur, *Haldanodon* already had acquired the mode of tooth replacement which characterizes the mammals sensu stricto

Many primitive characters are present in the lower jaws and the skull, indicating that *Haldanodon* represents a rather basal group of mammals. Thus, although *Haldanodon* already has a secondary jaw joint, i.e. the dentary (lower jaw bone) directly articulates with the squamosal, remains of the primary jaw joints are still present. They are reduced to small, dysfunctional bone fragments which were placed on the medial side of the jaw at the posterior end of the sulcus for MECKEL's cartilage (Fig. 15.4). Since they were only attached to the dentary by ligaments, they separated early during the decay of the animal, and are only represented by small grooves on the dentary. Since the bony elements of the primary jaw joint were still attached to the lower jaw, it can be concluded that *Haldanodon* did not have three, but only one bone in the middle ear cavity, as it is the case in reptiles. In mammalian evolution, the dysfunctional bones of the primary jaw joint were adapted to other functions and became part of the sound transmitting structures in the middle ear (malleus and incus). Consequently, LILLEGRAVEN & KRUSAT (1991) only found one bone in the middle ear (stapes) in their analysis of the skull remains of *Haldanodon*. Thus, *Haldanodon* lacks one of the essential characters of the mammals sensu stricto.

References

KERMACK, K. A., LEE, A. J., LEES, P.A., and MUSSETT, F. (1987): A new docodont from the Forest Marble. – Zoological Journal of the Linnean Society **89**: 1-39.

KOENIGSWALD, W. V. & STORCH, G. (1983): *Pholidocercus hassiacus*, ein Amphilemuride aus dem Eozän der "Grube Messel" bei Darmstadt (Mammalia, Lipotyphla). – Senckenbergiana lethaea **64**: 447-495.

KRUSAT, G. (1980): *Haldanodon exspectatus* KÜHNE & KRUSAT 1972 (Mammalia, Docodonta). Contribuição para o Conhecimento da Fauna do Kimeridgiano da Mina de Lignito Guimarota (Leiria, Portugal) IV Parte; VIII. – Memórias dos Serviços geológicos de Portugal, (nova Sér.) **27**: 1-79.

– (1991): Functional morphology of *Haldanodon exspectatus* (Mammalia, Docodonta) from the Upper Jurassic of Portugal. – Fifth Symposium on Mesozoic Terrestrial Ecosystems and Biota, extended abstracts. Contributions of the Paleontological Museum, University of Oslo **364**: 37-38.

KÜHNE, W. G. & KRUSAT, G. (1972): Legalisierung des Taxon *Haldanodon* (Mammalia, Docodonta). – Neues Jahrbuch für Geologie und Paläontologie, Monatshefte **1972**: 300-302.

LILLEGRAVEN, J. A. & KRUSAT, G. (1991): Cranio-mandibular anatomy of *Haldanodon exspectatus* (Docodonta; Mammalia) from the Late Jurassic of Portugal and its implications to the evolution of mammalian characters. – Contributions to Geology, University of Wyoming **28**: 39-138.

WALDMAN, M. & SAVAGE, R. J. G. (1972): The first Jurassic mammal from Scotland. – Journal of the Geological Society **128**: 119-125.

15 Multituberculates from the Guimarota mine

GERHARD HAHN & RENATE HAHN

Multituberculates are small animals, usually of the size of a mouse or a rat. They are the longest lived group amongst the known mammals, with a lifetime as a group from the Late Jurassic to the Early Tertiary (Oligocene) (GRANGER & SIMPSON 1929), a time span of more than 100 million years. Within Mesozoic mammal communities, they occupied the ecological niche that is filled by the rodents today: they are small, omnivorous to herbivorous, exhibit a significant variety, and occur in high numbers of individuals. If multituberculates are found in a fossil locality, they are usually the most common group of mammals (HAHN 1978a).

With very few exceptions, the multituberculates from the Guimarota mine all represent a single family, the Paulchoffatiidae HAHN 1969. They are the oldest known multituberculates and thus of special importance for our knowledge of this group. In contrast to the pantotheres and docodonts, no remains of the postcranial skeleton have been found so far; only skull remains, lower jaws and isolated teeth are present. Their preservation is usually rather poor. All skull remains are strongly compressed and broken. The lower jaws also exhibit many breaks, and some parts of the bones might be offset from each other, thus giving the appearance of a shape that is not real.

The morphology of the skull (Figs. 15.1a-c)

As it is the case in many primitive mammals, the skull of the Paulchoffatiidae is low and elongate in lateral view (Fig. 15.1b): the braincase is not significantly higher than the long snout. The eye socket is very large and confluent with the temporal fenestra posteriorly, and the jaw joint is placed far posteriorly. The lower jaw is massive, exhibits a high coronoid process and lacks an angular process (a posteriorly directed spur at the posterior end of the lower jaw, which is typical for all higher mammals). Further important, but less striking characters are the following:

- The presence of a large lacrimal that is visible in lateral view (at the anterior rim of the eye socket).

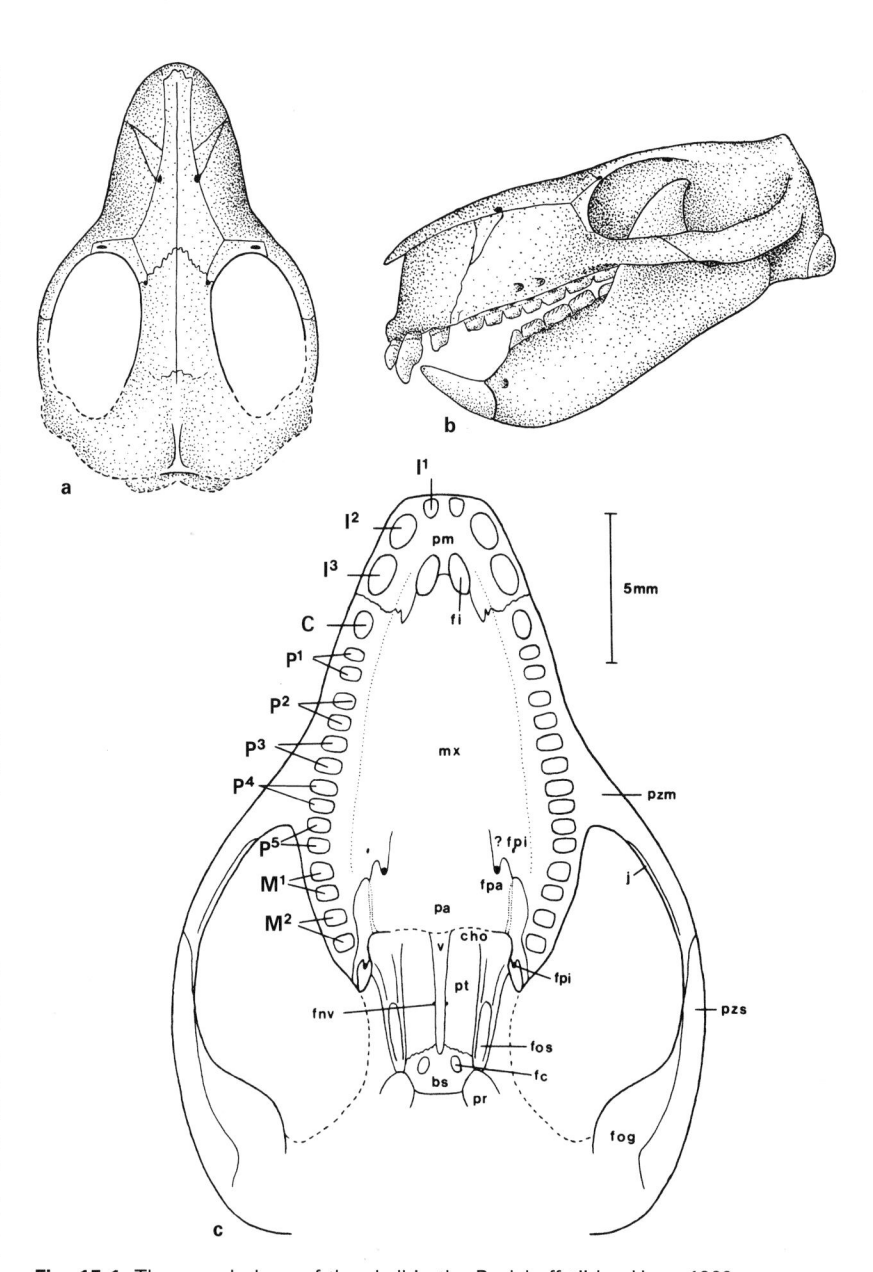

Fig. 15.1. The morphology of the skull in the Paulchoffatiidae HAHN 1969.
a. Skull of *Pseudobolodon* sp. in dorsal view. From HAHN & HAHN (1994: Fig. 2a).
b. Lateral view of *Pseudobolodon* sp. From HAHN & HAHN (1994: Fig. 2b).
c. Palatal view of an undetermined specimen (Paulchoffatiinae, gen. et sp. indet.). From (HAHN 1987: Fig. 4).

Fig. 15.2. Fragment of an anterior part of a skull of *Kuehneodon simpsoni* (V.J. 112-155, holotype) with dentition in lateral view. Length of the fragment 2.1 cm.

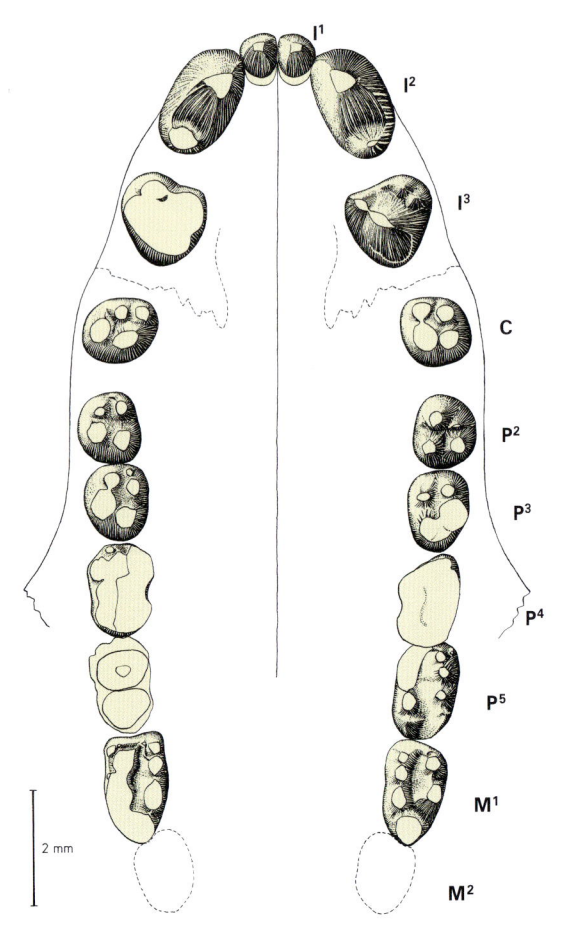

Fig. 15.3. The dentition of the upper jaw in *Kuehneodon simpsoni* HAHN 1969 (V.J. 112-155, holotype). The M^2 is missing. Occlusal view. From HAHN (1977: Fig. 4).

- The reduction of the jugal. In lateral view, the jugal arch is only formed by a posteriorly directed process of the maxilla and an anteriorly directed process of the squamosal. The jugal is situated as a thin splint on the inner side of the jugal arch and is not visible in lateral view.
- A small, triangular bone is placed between the premaxilla and the maxilla; this bone is interpreted as a septomaxilla. This is a bone that is typical for lower tetrapods (amphibians, reptiles), but is usually absent in mammals.
- The infraorbital foramen (the two small openings at the lower margin of the upper jaw, in front of the eye socket) is very small and simple, or subdivided into two openings. This is a primitive character, the implications of which will be discussed below (under "lifestyle").
- A small, rudimentary coronoid is present on the inner side of the lower jaw (not figured). This is again a bone that is typically found in reptiles, but is usually absent in mammals.

In dorsal view (Fig. 15.1a), the skull is characterized by the large, far posteriorly extending nasals (up to between the eye sockets), the also large lacrimals that are visible on either side at the anterior rim of the eye socket, and the large eye sockets, which are not separated from the temporal openings. The shortness of the posterior part of the skull is conspicuous, although not all details of this region are known yet.

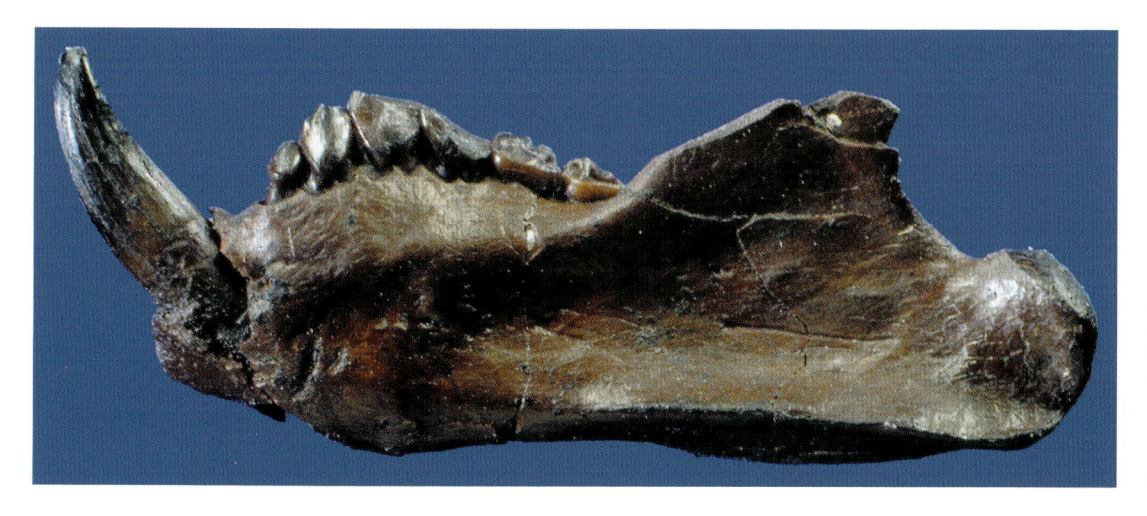

Fig. 15.4. Left lower jaw of *Meketibolodon robustus* (Gui Mam 89/76) in lateral view. The elongate incisor is well visible, which is separated from the cheek teeth by a gap. Length of the jaw 2.4 cm.

In palatal view (Fig. 15.1c), a plain palate is visible in the snout region. Large parts of it are formed by the maxillae (mx). A pair of small palatal openings (fi) (not present in all specimens) are found anteriorly on the border between the maxilla (mx) and premaxilla (pm), in the area of JACOBSON's organ. The region behind the internal narial openings (choana = ch) exhibits a complicated structure. It is divided into five parts and consists of a raised median ridge (v), two lateral ridges (fos), and two grooves in between (pt). This morphology is characteristic for all multituberculates. The jugal (j) is visible as a small splint on the inner side of the jugal arch on each side. The articular groove for the lower jaw (fog) is large, elongate and shallow. In the inner ear (not visible), the lagena (a part of the labyrinth) does not form a spiral, as it is the case in modern mammals, but it is straight, as in reptiles and also still in monotremes (platypus and echidnas).

The dentition (Figs. 15.2-15.6)

As in all other multituberculates, the teeth of the Paulchoffatiidae are multicuspid, "multituberculate".

In the upper jaws (Figs. 15.2 and 15.3), a maximum of 3 incisors (I), 1 canine (C), 5 premolars (P), and 2 molars (M) are found on each side. In comparison with the upper jaw, the number of teeth in the lower jaw (Figs. 15.4 and 15.5) is reduced. A maximum of 1I, 4P and 2M is present in each half of the jaw.

Only the first pair of the incisors in the upper jaw (I^1) still has a single cusp. The I^2 have one large cusp anteriorly and one to several posterior cusps. The I^3 exhibit a transverse ridge, which consists of 2 cusps, and additionally 0-3 small cusps on the margin of the crown anterolaterally and posterolingually. The I^{1-2} extend far ventrally. Such a complicated structure of the incisors is not found in any other group of mammals. The canine and the anterior premolars (P^{1-3}) have 3-4 cusps on average; specimens with two or five cusps are also found. The posterior premolars (P^{4-5}) have 2-3 longitudinally arranged rows of cusps. The molars have two rows of cusps. The canine is only distinguished from the anterior premolars by the fact that it only has one root, while the premolars exhibit two roots. Canines with 3-4 cusps are only found in multituberculates. The canine and all cheek teeth (C, P and M) have low tooth crowns.

The incisor of the lower jaw (Figs. 15.4 and 15.5) is very different from the I of the upper jaw. It is long, conical in outline and has only one cusp.

Fig. 15.5. The dentition of the lower jaw of *Guimarotodon leiriensis* HAHN 1969 (V.J. 461-155). The M_2 is missing. **a.** Lateral view. **b.** Occlusal view. From HAHN (1987: Fig. 5a-b).

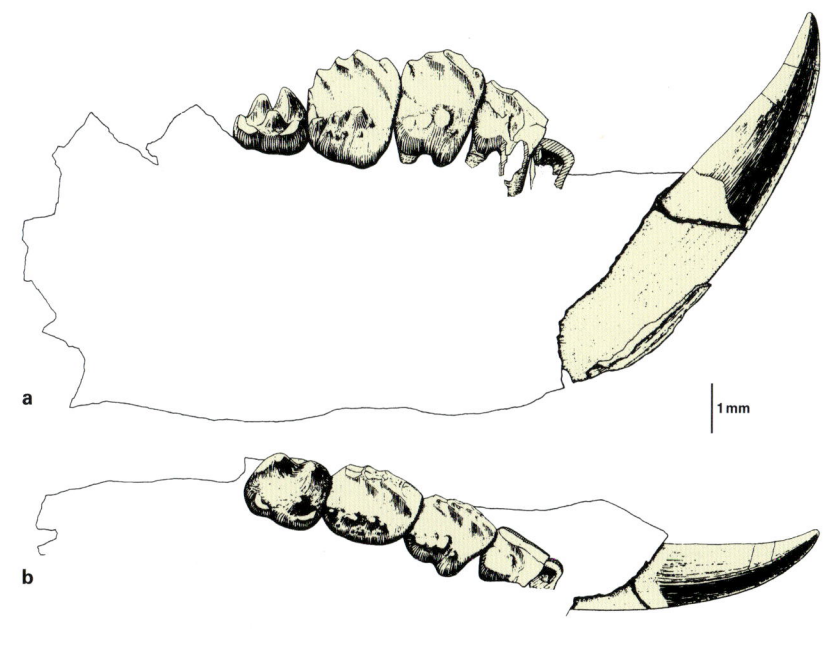

Fig. 15.6. The morphology of the M$_2$ in the Paulchoffatiidae HAHN 1969. Figured are two isolated specimens. From HAHN (1969: Fig. 36a, 37a).

Abb. 15.7. Wear in the premolars of the lower jaw, illustrated in a specimen of *Kuehneodon dietrichi* HAHN 1969. From HAHN (1978b). The premolars are completely worn down to the level of the molars. A functional difference between the two types of teeth is no longer present (V.J. 421-155).

It does not stand vertically in the jaw, but is attached to the latter in a more or less lying position. Its long root extends posteriorly to underneath the premolars, or even the molars and thus occupies a lot of space in the lower jaw. This results in the development of a short gap (diastema) between the incisor and the first premolar, and in a slightly oblique arrangement of the whole row of cheek teeth (anteriorly offset laterally, posteriorly offset lingually).

Principally, the premolars also have two rows of cusps, as it is the case in the P of the upper jaw. In the upper jaw, the two rows are of approximately equal height. In the lower jaw, however, the inner row of cusps is strongly enlarged and forms a cutting edge. In contrast, the outer row of cusps is very low and reduced to a row of "basal cusps". The premolars increase in length from anterior to posterior.

The M$_2$ exhibits a very unusual morphology (Figs. 15.6a and b). This tooth has only one cusp, which is situated in the anterior inner edge of the crown. The remaining area of the crown forms a groove, framed by a raised edge, in which the remains of other cusps are still visible. The anterior molar (M$_1$) has a normal morphology, with two rows of cusps, each of which has three cusps.

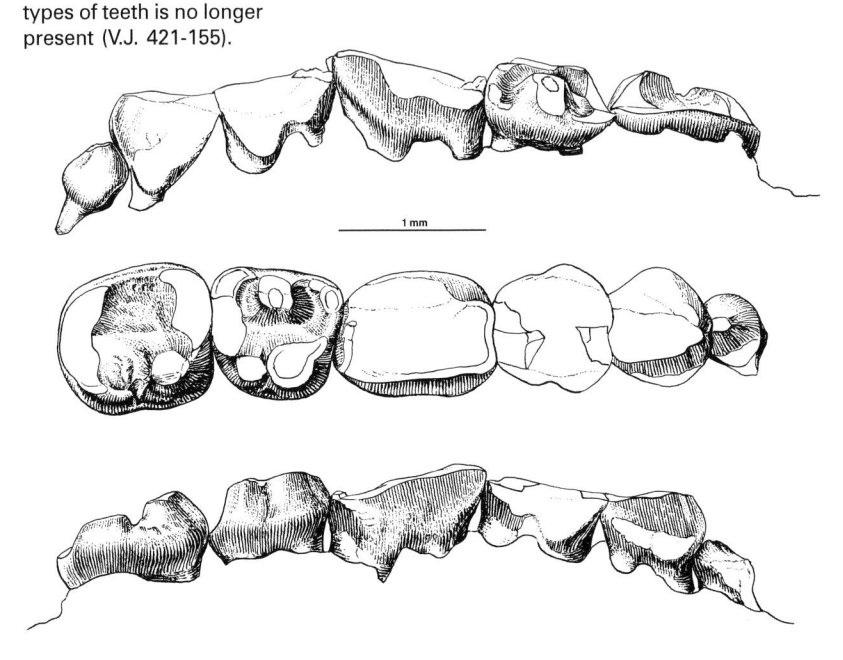

Biology (Figs. 15.7-15.10)

Sense organs: One of the most striking characters in the skull of the Paulchoffatiidae are the very large eye sockets, which are found in all preserved specimens. They indicate a nocturnal lifestyle of the animals. The sense of smell was probably also very well developed, since the olfactory lobes in the brain of multituberculates are the relatively largest known within mammals. No special statements can be made about the sense of hearing. It is usually well developed in nocturnal animals.

Dentition and feeding: Functionally, three different units can be distinguished in the dentition of the multituberculates: the food items are gathered with the incisors, then cut down between the premolars, which function as shears, and finally crushed between the molars. The lower jaw can be opened widely. The chewing motion is from posterior to anterior, not from anterior to posterior, as is the case in rodents. Such a partitioned ("plagiaulacoid") dentition with three functional units is also found in some fossil primates, some fossil marsupials from South America, and even today in the most basal kangaroos (*Bettongia*, rat kangaroo) in Australia.

The Paulchoffatiidae exhibit a more primitively functioning dentition than the geologically younger multituberculates. The premolars in the upper jaws are still broad and not transformed into shears in this group; a cutting function is not yet possible. The premolars of the lower jaw, which are "preadaptively" already transformed into shears, grind against the premolars of the upper jaw and are thus rapidly worn down. The result can be seen in Fig. 15.7: the premolars of the lower jaw are almost or completely worn down to the level of the molars. No difference in function exists between the premolars and the molars anymore. The food is ground down with the help of both types of teeth, cutting down the food items is not yet possible.

The wear of the teeth is very odd and difficult to interpret. In multituberculates, the opposed teeth of the upper and lower jaws do not act in exact combination, and the cusps of the opposed teeth are not interlocking if the jaws are brought into occlusion.

The number of cusps in identical teeth (for example the P^5 of the left and right side) is not always exactly the same. The row of molars is not worn down equally. The posterior premolars (P^{4-5}, see Fig. 15.3) are usually more strongly worn than the anterior premolars and the molars. The corresponding teeth of the right and left side can show significant differences in their wear. In Fig. 15.3, the right I^3 is so strongly worn down that the cusps are no longer distinguishable, while the left i^3 shows almost no signs of wear. In the molars, however, the right M^1 is more strongly worn down than the left M^1. Whereas in the latter, the individual cusps are clearly visible, the cusps of the outer row of the right M^1 already start to merge.

Similar patterns are found in the lower jaw. In the lower jaw figured in Fig. 15.8, the P_3 shows significant wear: its cutting edge is completely removed and its surface forms a concave groove. In contrast, the P_4 is almost unworn; the individual ridges of the cutting edge are still visible. The tooth replacement pattern might provide explanations for these unusual wear patterns. It is known with certainty for the I, C and P. As far as we know, the molars were not replaced, as it is the case in modern mammals. The tooth replacement pattern in multituberculates is still very primitive, similar to that of the mammal-like reptiles. The teeth are not replaced in one row, from anterior to posterior, but alternating (Hahn & Hahn 1998a). After the replacement of the P^1, that of the P^3 followed, than the P^5 was replaced, afterwards the P^2 and finally the P^4. Such a tooth replacement pattern is helpful to explain the present wear pattern. In the lower jaw illustrated in Fig. 15.8, the P_4 has already been replaced and thus does not show significant wear, whereas the P_3 has not yet been replaced and therefore exhibits much stronger wear. Consequently, it can be assumed that the left I^3 has already been replaced in the upper jaw figured in Fig. 15.3, but not the right I^3. If the teeth were replaced repeatedly, as it is the case in reptiles, or if they were already only replaced once, as in mammals (milk teeth – permanent teeth), is still unknown. However, a repeated replacement of the I, C and P in combination molars that are not replaced at all would be very unusual; such a tooth replacement pattern is otherwise unknown among mammals.

What did the animals eat? The very strong wear of the molars (see Fig. 15.7) indicates relatively hard food items. This conclusion is supported by the very well developed jaw musculature (well developed attachment areas for the jaw musculature and a high coronoid process in the lower jaw). Insects (e.g. beetles with hard wing covers) and parts of plants are possible food items. Possible plant food includes the fruits of ginkgos and the seams of conifers. The benettites were also widely distributed, so that their seams are also a possible food source. The Paulchoffatiidae could not gnaw their food as Recent rodents do. Their incisors are covered by enamel on all sides and they had determinate growth. Since their dentition could furthermore also not be used for cutting, the animals were certainly not able to cut up larger food items, such as roots. Their food must have been available in "mouth-sized" pieces, which could then be picked up with the help of the incisors. This is the case with the insects and the seams of the plants mentioned above. A movable tongue was probably present to help manipulating the food. This is indicated by the

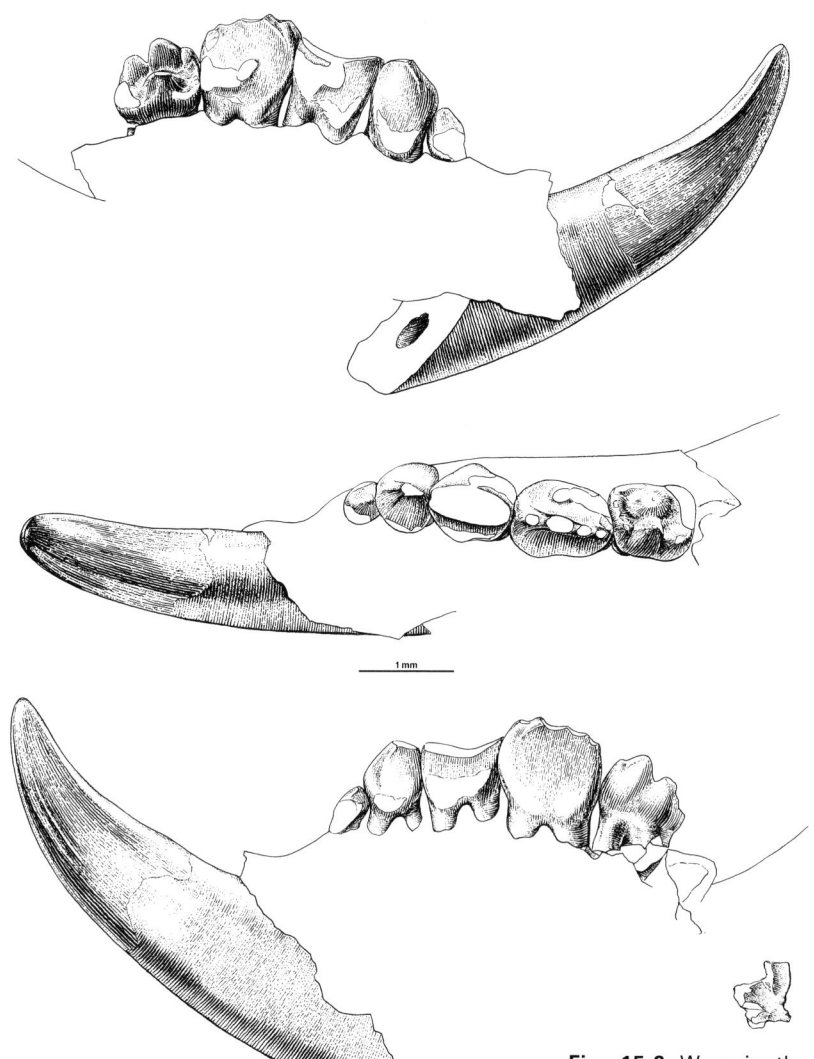

Fig. 15.8. Wear in the premolars of the lower jaw, illustrated in a specimen of *Kuehneodon dietrichi* Hahn 1969. From Hahn (1978b). The P_3 is strongly worn and shows a concave wear facet; in contrast, the P_4 exhibits barely any wear at all (V.J. 422-155).

Fig. 15.9 and 15.10. Reconstructions of multituberculates.
15.9. *Nemegtbaatar gobiensis* KIELAN-JAWOROWSKA 1974. The animal is illustrated in walking position, with widely splayed limbs. From KIELAN-JAWOROWSKA & GAMBARYAN (1994: Fig. 61).

dentition patterns in geologically younger multituberculates. In these animals, the anterior premolars of the upper jaws have no opposed teeth in the lower jaw. Since they are nevertheless still present, they must at least partially have worked in combination with the tongue. Thus, the Paulchoffatiidae can be interpreted as being omnivorous, with a tendency towards herbivory.

Lifestyle: The postcranial skeleton of the Paulchoffatiidae is unknown. Therefore, no statements can be made about their possible locomotion and thus their exact habitat. However, some considerations can be made, based on geologically younger multituberculates, of which the complete skeleton is known. One of these animals is *Nemegtbaatar* KIELAN-JAWOROWSKA 1974 from the Late Cretaceous of Mongolia (Fig. 15.9). KIELAN-JAWOROWSKA & GAMBARYAN (1994) published a reconstruction of the animal. The widely splayed out limbs are striking; according to the morphology of the skeleton, the possibility to move them in a parasaggital plane underneath the body was more restricted than in modern mammals. Fast running was probably only possible at short distances and with the help of short, high jumps. The usual locomotion was probably rather slow. *Nemegtbaatar* was presumably a nocturnal herbivore, which lived in an open landscape and was also able to dig. According to KIELAN-JAWOROWSKA & GAMBARYAN (1994), this taxon was to a certain degree similar to the Recent gerbils (Gerbillinae).

The interpretation of the lifestyle of *Ptilodus* COPE 1881 from the Paleocene of North America by KRAUSE & JENKINS (1983) is completely different. They concluded that these animals were swift climbers, which lived in trees, similar to squirrels. Indicators for such a lifestyle are, amongst others, the extendable thumb, the very flexible tarsus, and a differentiation of the last caudal vertebrae, which indicates the presence of a grasping tail.

So which of these two possibilities described above is the right one? Probably both! If one considers how many different lifestyles are found in Recent rodents (climber, runner, burrower, swimmer), a similar ecological differentiation can also be expected for the multituberculates. Unfortunately, however, this insight is of course not very helpful for the determination of the lifestyle of the Paulchoffatiidae. The specimens found in the Guimarota mine have been washed into this environment, their original habitat remains unknown. They might have lived on the ground in the open hinterland (together with the burrowing docodonts), but as well also on the trees close to the water. However, it seems very probable that the Paulchoffatiidae had a very gracile skeleton (which would be an argument against burrowing habits). In contrast to the eupantotheres and the Docodonta, not a single element of their postcranial skeleton has been found so far, which, given the high amount of material, can not be chance alone.

Life expectancy and enemies: The life expectancy of the Paulchoffatiidae was probably not very high. This is indicated by the primitive morphology of the dentition. The molars had (in contrast to most rodents) low tooth crowns; the cusps were (again in contrast to the rodents) not bound together in ridges, and the tooth enamel (again in contrast to the rodents) remained in a preprismatic condition (SANDER 1997). Such a dentition is rapidly worn down, and only the tooth replacement prolongs its duration. Thus, it can be assumed that the animals only lived a few years, as it is the case in many small rodents today (despite the better developed dentition).

The Paulchoffatiidae did not have to fear any enemies among their contemporaneous mammals, since representatives of the only carnivorous group of mammals from the Mesozoic, the Triconodonta, have not been found in the Guimarota mine so far. Birds of prey had not yet evolved. Thus, only reptiles can be regarded as possible enemies, especially small dinosaurs. As far as we know, the latter were diurnal, which explains the nocturnal lifestyle of the Paulchoffatiidae.

Reproduction: Today, mammals exhibit three different types of reproduction: the most basal is the reproduction by eggs (Monotremata), without any connection between the mother and the embryo. Only a loose connection between mother and embryo is present in marsupials. The juveniles are very small and are being born in a very undeveloped state; their further ontogenetic development takes place in the pouch. Finally, the "true" mammals with a placenta have the widest distribution today.

Which of these three types of reproduction was present in the multituberculates can again only be analyzed in geologically younger representatives of this group, which come from the Cretaceous and the Early Tertiary. The most detailed studies were again carried out by KIELAN-JAWOROWSKA & GAMBARYAN (1994) on the basis of taxa from the Late Cretaceous of Mongolia. The most conspicuous result of this investigation was that the pelvis in multituberculates was extraordinarily narrow, relatively more narrow than in all Recent mammals. This led to the question if it is possible to lay eggs through such a narrow birth tract at all. KIELAN-JAWOROWSKA & GAMBARYAN (1994) considered this to be rather improbable. Other paleontologists noted that the eggs might have been small and somewhat compressible (similar to the situation found in Recent turtles), which would then allow a reproduction by eggs. "Premature births", as they are found in marsupials, would have been able to pass the narrow birth tract of the multituberculates without any problems in any case, and thus, a reproduction strategy similar to that of the marsupials can at least not be excluded. Marsupial bones were present in multituberculates. However, since these elements obviously occurred in all mammals at the beginning of their evolution und probably served for the attachment of hind limb muscles, their presence does not help to answer the question at hand. Thus, we do not know: did the Paulchoffatiidae lay eggs, or did they raise their young in a pouch, similar to the Recent marsupials?

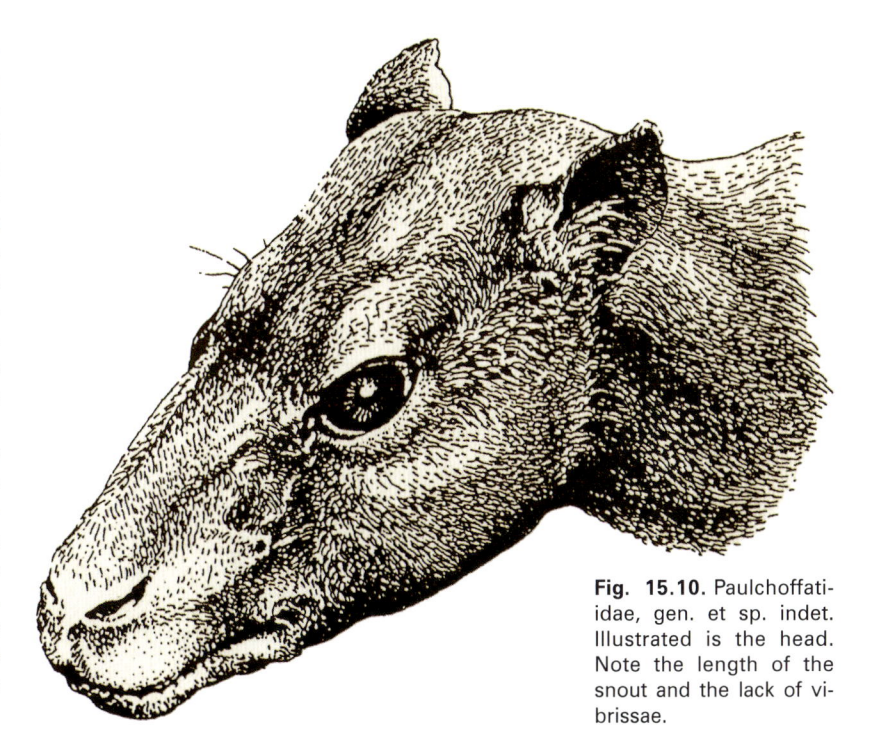

Fig. 15.10. Paulchoffatiidae, gen. et sp. indet. Illustrated is the head. Note the length of the snout and the lack of vibrissae.

Appearance

Fig. 15.10 shows a reconstruction of the head of a paulchoffatiid. The snout is figured longer and more slender than in *Nemegtbaatar* KIELAN-JAWOROWSKA 1974 (Fig. 15.9) and *Ptilodus* COPE 1881, which reflects the osteology. However, some details are of course subject to interpretation.

One of these is the shape of the ears. The illustration of the outer ear is very similar in all three reconstructions; it is based on the morphology found in many Recent small mammals.

But what did the pupils look like? Recent mammals show three different morphologies: round pupils, vertical slit-like pupils, and horizontal slit-

Fig. 15.11. The premolars (and canines) of the genera of the Paulchoffatiinae HAHN 1971. The left tooth row is illustrated. Hatched areas are worn and have lost their cusps. From HAHN & HAHN (1994: Fig. 5).
a. *Meketichoffatia krausei* HAHN 1993 (V.J. 110-155). The canine is preserved. Only the P^5 has three rows of cusps.
b. *Pseudobolodon oreas* HAHN 1977 (V.J. 397-155). The canine is preserved; P$^{4-5}$ have three rows of cusps. The P$^{2-3}$ exhibit three to four cusps.
c. *Pseudobolodon krebsi* HAHN & HAHN 1994 (V.J. 447-155). In contrast to *Ps. oreas*, the P$^{2-3}$ have five cusps.
d. *Henkelodon naias* HAHN 1977 (V.J. 401-155). As in *Pseudobolodon*, the P$^{4-5}$ have three rows of cusps, but in contrast to this genus, the canine is missing.
e. *Kielanodon hopsoni* HAHN 1987 (V.J. 463-155). The P^3 has two rows of cusps, the P$^{4-5}$ have three rows of cusps each. C and P$^{1-2}$ are unknown.

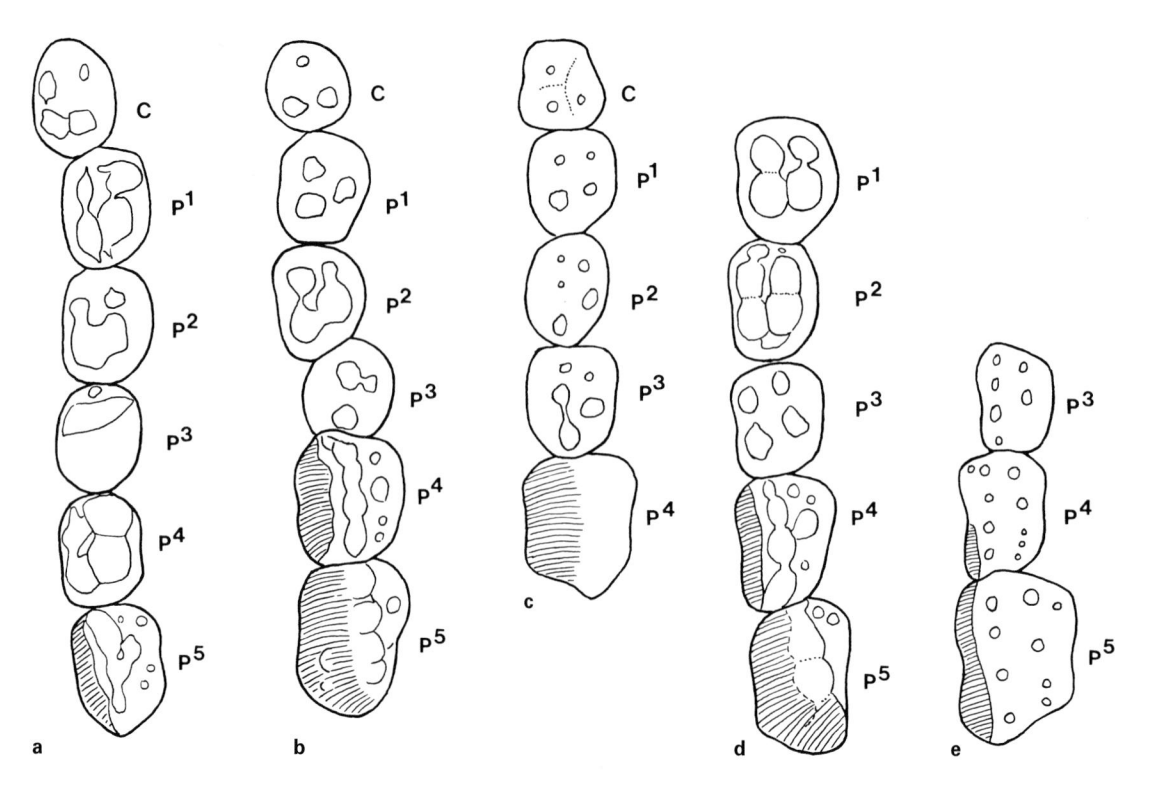

like pupils. The first shape is for example found in primates, the second in cats and the third in sheep and goats. Round pupils probably represent the most primitive shape, and thus they are figured as being round in the paulchoffatiid and *Nemegtbaatar*. However, *Ptilodus* shows a cat-like vertical slit.

A third region of uncertain reconstruction is the presence of vibrissae on the tip of the snout. Were they present, or not, and if they were, how long were they? Vibrissae have been illustrated for *Nemegtbaatar* and *Ptilodus*, but not in the paulchoffatiids. The reason for this is the development of the infraorbital foramen. As mentioned under "skull anatomy", the latter was very small in the paulchoffatiids. The supply of the snout with blood vessels and nerves was probably less well developed than in modern mammals and also geologically younger multituberculates. In reptiles, a correspondingly small infraorbital foramen is connected with an immovability of the snout region (lack of facial expression) and the lack of vibrissae. The lack of such hairs in the reconstruction of the head of the paulchoffatiid is thus meant to emphasize the primitive morphology of the snout in these animals.

Which size the head had, relative to the body, can again only be decided on the basis of geologically younger multituberculates. The relative size of the skull is strikingly larger in the reconstruction of *Nemegtbaatar* that in that of *Ptilodus*. The size ratios in the former genus are probably more accurate than those in the latter genus; they better match the proportion exhibited by the skeleton. The small skull, the shape of the pupil and the naked tail in *Ptilodus* remind of the Recent didelphids, which were – consciously or unconsciously – obviously used as a model for this reconstruction.

Systematics (Fig. 15.11)

Recording the species diversity is problematic because of the poor preservation, especially of the skull remains, and the fact that no skull has been found in association with a lower jaw. Of several hundred fossils, only 18 skull remains and 50 lower jaws could be unambiguously determined systematically so far (see Tab. 15.1). Since – with the exception of one genus – skulls and lower jaws cannot be correlated, a separate systematic determination must be proposed for both elements.

Five genera can be distinguished in the skull remains and four in the lower jaws. In addition, three genera are only known from isolated teeth (HAHN & HAHN 1998c,d) (see Tab. 15.1). Thus, it

Fig. 15.12. Evolutionary changes of the crown morphology of M_1 in Paulchoffatiidae. **a.** plesiomorph condition: cusp B_2 sits at the buccal margin of the crown. Cusps B_1 and B_2 are moderately large and not subdivided. This condition is present in *Kuehneodon dietrichi* HAHN 1969 and in *Meketibolodon robustus* (HAHN 1978). **b.** cusp B_2 is slightly shifted from the buccal margin towards the middle of the tooth crown. Cusps B_1 and B_3 are subdivided in 2-4 small cuspules. This condition characterizes m_1 of *Kuehneodon uniradiculatus* HAHN 1978. **c.** cusp B_2 is completely shifted to the center of the tooth crown. Cusps B_1 and B_3 are changed to a row of small cuspules along the buccal margin of the tooth crown. This condition is present in *Guimarotodon leiriensis* HAHN 1969.

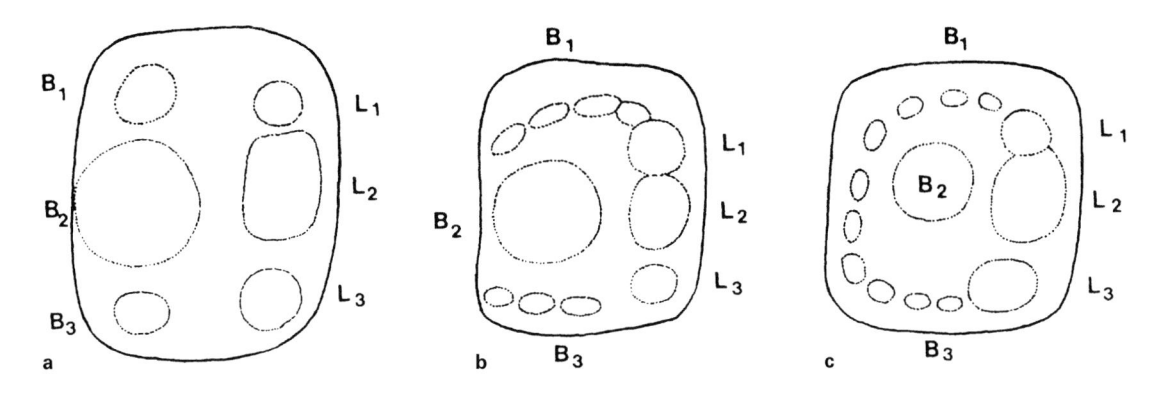

can be assumed that seven to eight genera of multituberculates lived in the surroundings of the Guimarota mine. This significant diversity becomes understandable if one considers that the animals have been washed into the coals of Guimarota, whereas they probably lived in different habitats during life. The taphocoenosis (grave community) does not correspond to the living association in the sense of a biocoenosis.

Because of the poor preservation, the systematic determination of the skull remains is only based on the dentitions. The following characters are important for the taxonomic determination: 1) number and arrangement of the cusps on the M^1 (see HAHN & HAHN 1998c: Figs. 1a-f). 2) The morphology of the P^{4-5}, especially the presence or absence of an additional third row of cusps. 3) The morphology of the P^3, which might be either similar to the P^{1-2} (short, rounded), or to the P^4 (elongate, with two rows of cusps). 4) The presence or absence of the canines (C). The morphology of the canine and the premolars in important species is illustrated in Fig. 15.11. It should be noted, however, that this pattern should already be modified on the basis of new results from the ongoing research on isolated teeth.

Apart from the dentition, the mandible itself can be used for the characterization of the genera represented by lower jaws (Tab. 15.2).

Paulchoffatia KÜHNE 1961 is characterized by a massive Corpus mandibulae (the part of the jaw below the tooth row), a rounded lower margin of the jaw and a massive, only slightly curved and steeply inclined incisor with a short root.

Meketibolodon HAHN 1993 is distinguished from the other genera by two characters: the tooth row is significantly convexly curved upwards, and the Corpus mandibulae has angled margins ventrally (HAHN & HAHN 1998b). The incisor is more strongly curved than it is the case in *Paulchoffatia*, and its root is longer. The Corpus mandibulae is similarly massive to that in *Paulchoffatia*.

Guimarotodon HAHN 1969 exhibits a more slender Corpus mandibulae than either *Paulchoffatia* or *Meketibolodon*. The most conspicuous character of this genus is the morphology of the P_{3-4} and the M_1. In contrast to the situation in the other genera, the P_{3-4} bear two rows of basal cusps on top of each other. In the M_1, the second, enlarged cusp of the outer row (B_2) is displaced from the rim of tooth towards the center of the crown, and it is surrounded by a circle of small cusps. The incisor is relatively little curved and its

Tab. 15.1. The taxa of Paulchoffatiidae HAHN 1969 found in the Guimarota mine, including the number of known specimens. – *Plesiochoffatia*, *Xenachoffatia* and *Bathmochoffatia* are only known from isolated teeth. *Plesiochoffatia* HAHN & HAHN 1999 was originally described under the name *Parachoffatia* HAHN & HAHN 1998c; however, the latter name represents a homonym of the perisphinctid ammonite *Parachoffatia* MANGOLD 1970.

Taxon	Lower jaw	Skull	isolated teeth M_2	M^1
Paulchoffatia delgadoi KÜHNE 1961	5			
Paulchoffatia sp. A HAHN 1978	4			
Meketibolodon robustus (HAHN 1978b)	9			
Guimarotodon leiriensis HAHN 1969	3			
Kuehneodon dietrichi HAHN 1969	20			
Kuehneodon uniradiculatus HAHN 1978b	5			
Kuehneodon guimarotensis HAHN 1969	4			
Meketichoffatia krausei HAHN 1993		2		
Pseudobolodon oreas HAHN 1977		7		
Pseudobolodon krebsi HAHN & HAHN 1994		2		
Henkelodon naias HAHN 1977		1		
Kielanodon hopsoni HAHN 1987		3		
Kuehneodon simpsoni HAHN 1969		1		
Kuehneodon dryas HAHN 1977		2		
Plesiochoffatia thoas (HAHN & HAHN 1998c)			3	
Plesiochoffatia staphylos (HAHN & HAHN 1998c)			1	
Plesiochoffatia peparethos (HAHN & HAHN 1998c)			1	
Xenachoffatia oinopion HAHN & HAHN 1998c			3	
Bathmochoffatia hapax HAHN & HAHN 1998c				1

root is of similar length as that of *Meketibolodon* and extends to underneath the posterior premolars.

Kuehneodon HAHN 1969 exhibits the most horizontally oriented attachment of the incisor and the most significant upward curve towards the tip of this tooth. It also has the longest root, which extends to underneath the molars. The Corpus mandibulae is similar to that of *Guimarotodon* in its slenderness. The P_{3-4} exhibit a normal morphology with only one row of basal cusps. In the M_1, one of the species, *K. dietrichi* HAHN 1969, shows the "normal" state with one large cusp B_2 close to the margin of the crown. In *K. uniradiculatus* HAHN 1978, however, the B_2 is already slightly displaced from the margin of the crown towards the medial part, but it still lacks the circle of small cusps found in *Guimarotodon* (Fig. 15.12). However, it is obvious that the morphology of the crown in *Guimarotodon* can be derived from that seen in *Kuehneodon*.

As mentioned above, a correlation between the lower jaws and the upper jaws is only possible for one genus, in *Kuehneodon*. Here, those upper and lower jaws are united, which exhibit the lowest number of derived characters (apomorphies), and are thus closest to the main evolutionary lineage of the multituberculates. Thus, *Kuehneodon* combines premolars that lack a third row of cusps in the upper jaw with a derived lower jaw, in which the attachment of the incisor is most horizontally directed and the tooth row shows the strongest diagonal orientation in relation to the longitudinal axis of the jaw (see Tab. 15.2). A separate subfamily, Kuehneodontinae HAHN 1971, was created for this advanced genus. It forms the counterpart to the subfamily Paulchoffatiinae HAHN 1971. The latter includes all other genera, and a correlation of the dentitions of the upper and lower jaws of its members is not possible.

The family Albionbaataridae is represented only by two isolated upper molars (M^1) in the Guimarota mine (HAHN & HAHN 1998d). They are smaller than the molars of the Paulchoffatiidae and have more cusps which are oriented in two rows. The cusps are pointed and are characterized by radially oriented enamel crests (Fig. 15.13). The family Albionbaataridae has been established at the base of isolated molars from the Purbeck of southern England (KIELAN-JAWOROWSKA & ENSOM 1994) and belongs to the Plagiaulacoidea. The teeth from the Guimarota mine are somewhat older and indicate that the Albionbaataridae probably are evolved from the Paulchoffatiidae.

Phylogenetic significance

The Paulchoffatiidae HAHN 1969 are the oldest family that can undoubtedly be referred to the Multituberculata. All other, geologically older groups – Haramyidae SIMPSON 1947 (Late Triassic), Theroteinidae SIGOGNEAU-RUSSELL, FRANK & HEMMERLÉ 1986 (Late Triassic) and Eleutherodontidae KERMACK, KERMACK, LEES & MILLS 1998 (Middle Jurassic) – might belong to the Allotheria MARSH 1880 as separate evolutionary lineages, but they do not represent the direct ancestors of the multituberculates. Since mainly isolated teeth are known of all of the three groups mentioned, many questions concerning their systematic position remain unanswered at present.

As the oldest certain members of the Multituberculata, the Paulchoffatiidae are thus of special

Tab. 15.2. The characters in the lower jaws of the Paulchoffatiidae HAHN 1969.

Character		*Paulchoffatia*	*Meketibolodon*	*Guimarotodon*	*Kuehneodon*
Incisors	Crown outline	massive, straight	slender, curved	slender, straight	slender, strongly curved
	Orientation	steeply inclined	intermediate	intermediate	horizontal
	Length of the root	below P_{1-2}	below P_{3-4}	below P_{3-4}	below M
	Flexure of the tooth row	slightly convex	significantly convex	straight	straight
Cheek teeth	Angle between tooth row and jaw	7°-10°	12°-16°	15°	20°
	Roots P_1	2	1	2	1-2
	Basal cusps P_4	1 row	1 row	2 rows	1 row
	Position of B_2 on M_1	marginal	marginal	offset towards the crown center	marginal
Mandible	Corpus mandibulae	massive	massive	slender	slender
	Ventral margin	rounded	angled	rounded	rounded
	Length (mm)	22-28	23-25	20-22	20-26

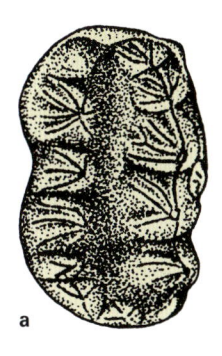

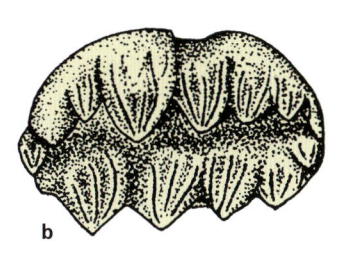

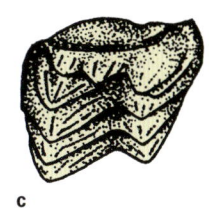

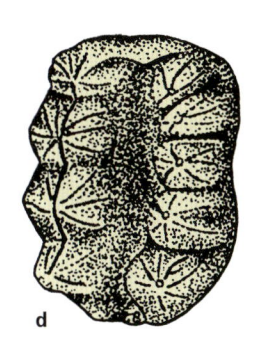

Fig. 15.13. Upper molars (M^1) of *Proalbionbaatar plagiocyrtus* HAHN & HAHN 1998d. **a-c.** left molar (IPFUB pl (M1)-1 sin., holotype) in occlusal (**a**), oblique lingual (**b**), and distal (**c**) view. **d.** right molar (IPFUB pl (M1)-2 dext., paratype) in occlusal view. Scale bar = 1 mm. From HAHN & HAHN 1998d)

interest for the evolution of this order. The question is, in how far they can be regarded as the ancestral model of the geologically younger multituberculates in terms of their morphology, and in how far they exhibit derived characters that exclude them from being direct ancestors. In respect to the anatomy of the skull, the Paulchoffatiidae are more primitive than younger multituberculates in almost all characters. Thus, in summary it can be said that they generally conform to the expected ancestral morphology of the younger multituberculates in respect to the structure of the skull and the dentition. However, this is not the case for at least one character, the morphology of the M$_2$. Therefore, another basal group of multituberculates, which has not been discovered yet, must have been present in the early Late Jurassic. This group was probably similar to the Paulchoffatiidae (especially the Kuehneodontinae) in respect to their general anatomy, but did not have specialized M$_2$ with two rows of cusps. This consideration does not lessen the significance of the Paulchoffatiidae for the evolution of all multituberculates. They really exhibit the basal multituberculate bauplan in many characters, without which every discussion about the evolution of this group would be incomplete.

References

COPE, E. D. (1881): Eocene Plagiaulacidae. – American Naturalist **15**: 921-922.
GRANGER, W. & SIMPSON, G. G. (1929): A revision of the Tertiary Multituberculata. – Bulletin of the American Museum of Natural History **56**: 601-676.
HAHN, G. (1969): Beiträge zur Fauna der Grube Guimarota Nr. 3. Die Multituberculata. – Palaeontographica A **133**: 1-100.
– (1971): The dentition of the Paulchoffatiidae (Multituberculata, Upper Jurassic). – Memória dos Serviços Geológicos de Portugal, n. s. **17**: 7-39.
– (1977): Neue Schädel-Reste von Multituberculaten (Mamm.) aus dem Malm Portugals. – Geologica et Palaeontologica **11**: 161-186.
– (1978a): Die Multituberculata, eine fossile Säugetier-Ordnung. – Sonderband des naturwissenschaftlichen Vereins Hamburg **3**: 61-95.
– (1978b): Neue Unterkiefer von Multituberculaten aus dem Malm Portugals. – Geologica et Palaeontologica **12**: 177-212.
– (1987): Neue Beobachtungen zum Schädel- und Gebiss-Bau der Paulchoffatiidae (Multituberculata, Ober-Jura). – Palaeovertebrata **17**: 155-196.
– (1993): The systematic arrangement of the Paulchoffatiidae (Multituberculata) revisited. – Geologica et Palaeontologica **27**: 201-214.
HAHN, G. & HAHN, R. (1994): Nachweis des Septomaxillare bei *Pseudobolodon krebsi* n. sp. (Multituberculata) aus dem Malm Portugals. – Berliner geowissenschaftliche Abhandlungen E **13**, Festschrift Bernard KREBS: 9-29.
– (1998a): Neue Beobachtungen an Plagiaulacoidea (Multituberculata) des Ober-Juras. – 1. Zum Zahn-Wechsel bei *Kielanodon*. – Berliner geowissenschaftliche Abhandlungen E **28**: 1-7.
– (1998b): Neue Beobachtungen an Plagiaulacoidea (Multituberculata) des Ober-Juras. – 2. Zum Bau des Unterkiefers und des Gebisses bei *Meketibolodon* und bei *Guimarotodon*. – Berliner geowissenschaftliche Abhandlungen E **28**: 9-37.
– (1998c): Neue Beobachtungen an Plagiaulacoidea (Multituberculata) des Ober-Juras. – 3. Der Bau der Molaren bei den Paulchoffatiidae. – Berliner geowissenschaftliche Abhandlungen E **28**: 39-84.
– (1998d): Neue Beobachtungen an Plagiaulacoidea (Multituberculata) des Ober-Juras. – 4. Ein Vertreter der Albionbaataridae im Lusitanien Portugals. – Berliner geowissenschaftliche Abhandlungen E **28**: 85-89.
– (1999): Nomenklatorische Notiz: Namens-Änderung bei Multituberculata (Mammalia). – Geologica et Palaeontologica **33**: 156.
KERMACK, K. A., KERMACK, D. M., LEES, P. M. & MILLS, J. R. E. (1998): New multituberculate-like teeth from the Middle Jurassic of England. – Acta Palaeontologica Polonica **43**: 581-606.
KIELAN-JAWOROWSKA, Z. (1974): Results of the Polish-Mongolian palaeontological expeditions – Part V: Multituberculate succession in the Late Cretaceous of the Gobi Desert (Mongolia). – Palaeontologia Polonica **30**: 23-44.
KIELAN-JAWOROWSKA, Z. & ENSOM, P. C. (1994): Tiny plagiaulacoid multituberculate mammals from the

Purbeck Limestone Formation of Dorset, England. – Palaeotology **37**: 17-31.

KIELAN-JAWOROWSKA, Z. & GAMBARYAN, P. P. (1994): Postcranial anatomy and habits of Asian multituberculate mammals. – Fossils and Strata **36**: 1-92.

KRAUSE, D. W. & JENKINS, F. A. (1983): The postcranial skeleton of North American multituberculates. – Bulletin of the Museum of Comparative Zoology **150**: 199-246.

KÜHNE, W. G. (1961): Eine Mammaliafauna aus dem Kimmeridge Portugals. – Neues Jahrbuch für Geologie und Paläontologie, Monatshefte **1961**: 374-381.

MANGOLD, C. (1970): Les Perisphinctidae (Ammonitina) du Jura méridional au Bathonien et au Callovien. – Documents des Laboratoires de Géologie de la Faculté des Sciences de Lyon **41**: 1-246.

MARSH, O. C. (1880): Notice of Jurassic mammals representing two new orders. – American Journal of Science, 3 series **20**: 235-239.

SANDER, P. M. (1997): Non-mammalian synapsid enamel and the origin of mammalian enamel prisms: the bottom-up perspective. – In: KOENIGSWALD, W. V. & SANDER, P. M. [eds.] Tooth enamel microstructure: 41-62; Rotterdam (A. A. Balkema).

SIGOGNEAU-RUSSELL, D., FRANK, R. M. & HEMMERLÉ, J. (1986): A new family of mammals from the lower part of the French Rhaetic. – In: PADIAN, K. [ed.] The beginning of the age of dinosaurs: 99-108, Cambridge/Mass. (Cambridge University Press).

SIMPSON, G. G. (1947): *Haramiya*, new name, replacing *Microcleptes* SIMPSON, 1928. – Journal of Paleontology **21**: 497.

The dryolestids and the primitive "peramurid" from the Guimarota mine

Thomas Martin

The eupantotheres are the most common mammals in the Guimarota mine, with approximately 500 jaws and skull remains and one almost complete skeleton preserved. They are represented by two families in this locality, the Dryolestidae, and the Paurodontidae (including henkelotheriids). The eupantotheres are one of the ancestral groups of modern mammals (marsupials and placentals), but they are still on a much more basal evolutionary level, which becomes especially obvious in the structure of their teeth. An additional cusp (protocone) is developed in the molars of the upper jaw in modern mammals, which fits into a basin-like structure (talonid) in the molars of the lower jaw (Fig. 16.1), like a pestle into the mortar. This cusp is still absent in the eupantotheres. Here, the talonid does not form a basin-like structure, but is only developed as a little cusp. Therefore, the eupantotheres are also known as pre-tribosphenic mammals, in contrast to the tribosphenic mammals. The word tribosphenic means "grinding-wedge-like", which refers to the grinding-crushing function of the molars in marsupials and placental mammals.

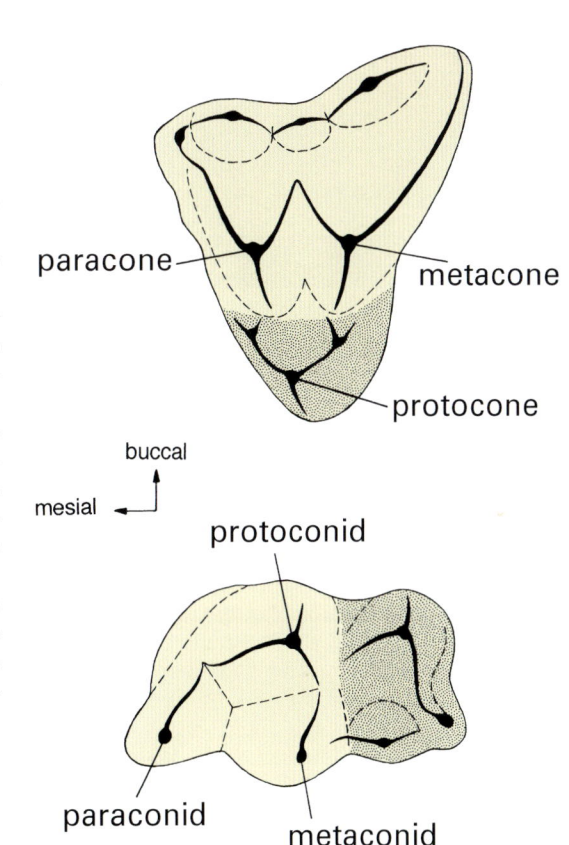

Fig. 16.1. Generalized pattern of an upper and lower tribosphenic molar. The protocone (stippled area in the upper molar) and the talonid (stippled area in the lower molar) are not developed in pre-tribosphenic molars. Redrawn and modified from Thenius (1989).

Dryolestidae

Within the eupantotheres, the dryolestids represent a clearly definable group that is characterized by their molars, which are significantly shortened in the longitudinal axis of the jaw (mesio-distally) and considerably broadened perpendicular to the jaw (labio-lingually). This character distinguishes the dryolestids from all other eupantotheres. A further striking character in the lower molars of the dryolestids is the discrepancy in size between the roots, with the anterior one being considerably larger than the posterior one. Looking at a lower jaw in lateral view, one could get the impression that the molars had only one root, since the small posterior root is completely obscured by the enormous anterior one (Fig. 16.2). Another important character for the family Dryolestidae is their high number of molars. Whereas the placental mammals have three molars and the marsupials have four molars in each side of the jaw, dryolestids have 7-8, sometimes even nine. This is not necessarily an especially primitive condition, as it might be assumed in comparison with reptiles, which have a high number of homogeneous teeth. In the ancestors of the dryolestids from the Early and Middle Jurassic (e.g. *Kuehneotherium*), the number of molars was reduced to 3-5. This number was secondarily increased again in the Dryolestidae, probably in connection with the shortening of the single molars. On the other hand, the talonid, the cusp at the posterior end of the molars of the lower jaw, which becomes a basin-like structure in the tribosphenic mammals, as mentioned above, is significantly larger in *Amphitherium* (Middle Jurassic) than in the dryo-

Fig. 16.2. Left lower jaw of *Dryolestes leiriensis* in lateral view (Gui Mam 41/79). Only the incisors and some molars have been lost in this superbly preserved specimen. To the right, the large coronoid process can be seen, which served as an attachment area for the jaw muscles. The process in the middle of the posterior end is the Processus articularis, which formed the joint with the skull. The small, rod-like process at the lower end of the jaw is the angular process, which is present in all modern mammals and the representatives of their stem lineage, but is missing in docodonts and multituberculates. Length of the jaw 3.2 cm.

lestids. Thus, the latter obviously represent a different evolutionary lineage from the one that led to the modern mammals.

Distribution of the Dryolestidae

The Dryolestidae are the longest-lived group of eupantotheres. The oldest find of dryolestids consists of two isolated molars of the lower jaw from the Middle Jurassic (Late Bathonian) Forest Marble of Oxfordshire/England (FREEMAN 1976). The next youngest, and at the same time best represented Dryolestidae are those from the Kimmeridgian of the Guimarota mine (MARTIN 1995, 1997, 1999). The classical dryolestid fauna from the Morrison Formation of Wyoming (MARSH 1878, 1879, SIMPSON 1929, PROTHERO 1981) is slightly younger, of Late Kimmeridgian age according to new micropaleontological data (SCHUDACK 1993, 1995, 1996). Apart from some jaw remains (SIMPSON 1928), several localities from the latest Jurassic and the Jurassic-Cretaceous boundary in southern England and Portugal (Porto Pinheiro) have mainly yielded isolated teeth of dryolestids (CLEMENS & LEES 1971, MARTIN 1999). The youngest known representative of the Dryolestidae from the Northern Hemisphere is *Crusafontia cuencana* HENKEL & KREBS 1969 from the lignite coals (Barremian) of Uña (Province of Cuenca) and the fluvio-lacustrine deposits of Galve in the province of Teruel (both Spain). After that, remains of this family are missing from the northern continents, with the exception of a fragmentary molar of a lower jaw from the Late Cretaceous "Mesaverde" Formation of Wyoming, which was referred to the Dryolestidae by LILLEGRAVEN & MCKENNA (1986). The Dryolestidae obviously survived until the Late Cretaceous in South America, where they experienced a significant radiation in the geographic isolation of this continent (BONAPARTE 1986, 1990, 1994).

Dryolestes leiriensis MARTIN 1999, the large dryolestid from the Guimarota mine

The most common dryolestid form the Guimarota mine is *Dryolestes leiriensis*, which is represented by 89 jaws and jaw fragments and several skull fragments (Figs. 16.2-16.4, 16.6). The genus *Dryolestes* was first described from the Morrison Formation of Wyoming; its presence in Portugal indicates close faunistic relationships between Europe and North America in the Late Jurassic, when both continents were still close together, and the northern Atlantic had just began to open.

Lower jaw (Figs. 16.2-16.5): With a length of the lower jaw of c. 3.5 cm, *Dryolestes leiriensis* was approximately the size of a hedgehog, which is quite large for a dryolestid. Four simple incisors are found in the lower jaw of *Dryolestes*; especially the anteriormost two of these teeth were oriented almost horizontally and protruded from the jaw (MARTIN 1999) (Fig. 16.4). This indicates that *Dryolestes* used the anteriormost part of its dentition like a comb to take care of its fur, as it is the case in many small mammals today. Behind the incisors, a large canine is present, which is securely implanted in the jaw with two roots (Figs. 16.4 and 16.5). Its crown consists of a large, triangular main

Fig. 16.3. Left lower jaw of *Dryolestes leiriensis* (Gui Mam 130/74, holotype) in inner (lingual) view. The dentition is very well preserved in this lower jaw, and only the incisors are missing. From right to left, the pointed canine, four premolars and eight (!) molars are visible. The shallow groove for MECKEL's cartilage can be seen at the ventral margin of the bone; it runs in a slight curve from the pterygoid fossa (left) towards the symphysis of the jaw (right). Preserved length of the jaw 3.0 cm.

tip, which is slightly recurved and thus especially suited for capturing prey items. Behind the canine, four premolars are found, which also have two roots each. The two anterior premolars are relatively small; the third and especially the fourth are considerably larger. The premolars are mainly distinguished from the canine by the presence of a small additional posterior cusp and their smaller size. They were mainly used for securing and killing the prey animal.

The molars exhibit a completely different morphology than the teeth of the anterior part of the dentition (Fig. 16.5). They consist of three main cusps that are arranged in the shape of a triangle. These cusps are called the proto-, para-, and metaconid (Fig. 16.1). The likewise triangular molars of the dentition of the upper jaw insert in between these triangles of the lower molars during chewing. During this process, the anterior cutting edge of each molar of the upper jaw slides along the posterior edge of the corresponding molar of the lower jaw and thus cuts the food items into pieces. At the posterior end of the molars of the lower jaw, a small cusp is found, which is called the talonid. As mentioned above, this cusp becomes the talonid basin in the modern mammals, in which the newly developed protocone of the molars of the upper jaw inserts. By this development, the crushing and grinding function of the molars is significantly enhanced in the so-called tribosphenic molars. The small size of the talonid indicates that the Dryolestidae do not represent the lineage that includes the direct ancestors of the modern mammals. Mammals with a significantly enlarged talonid already lived alongside the dryolestids. These mammals, which are known as "peramurids", are represented in the Guimarota mine by a primitive and very small form. However, this taxon is very rare, since only some four dozen teeth of it were found among the many thousand isolated mammal teeth resulting from the screen washing of the coals of Guimarota.

The high number of molars in the dryolestids seems to be particularly unusual. Whereas the

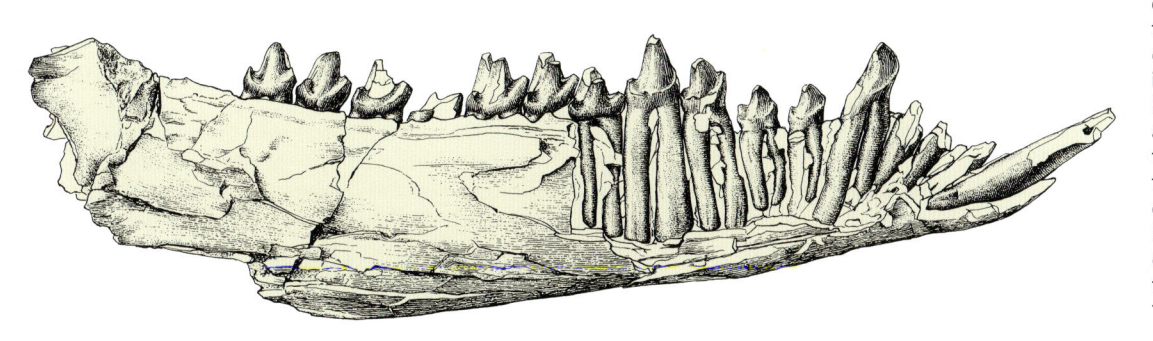

Fig. 16.4. Left lower jaw of *Dryolestes leiriensis*, in inner (lingual) view (Gui Mam 49/79). The broken bone allows a view of the enormous roots of the teeth of the anterior part of the dentition. The first incisor is almost oriented horizontally and was probably used to take care of the fur. The triangular attachment area for the coronoid is visible at the basis of the coronoid process. Preserved length of the jaw 2.8 cm. From MARTIN (1999).

Fig. 16.5. Tooth row of a left lower jaw of *Dryolestes leiriensis* (Gui Mam 130/74, holotype). **a** in dorsal (occlusal) view, **b** in inner (lingual) view, **c** in outer (labial) view. In **b**, the canine, four premolars (P_{1-4}), and eight molars (M_{1-8}) are visible from right to left; the incisors are not preserved. The canine and premolars have only one cusp, the molars consist of three large cusps and a smaller posterior one (talonid). Scale bar = 1 mm. From MARTIN (1999).

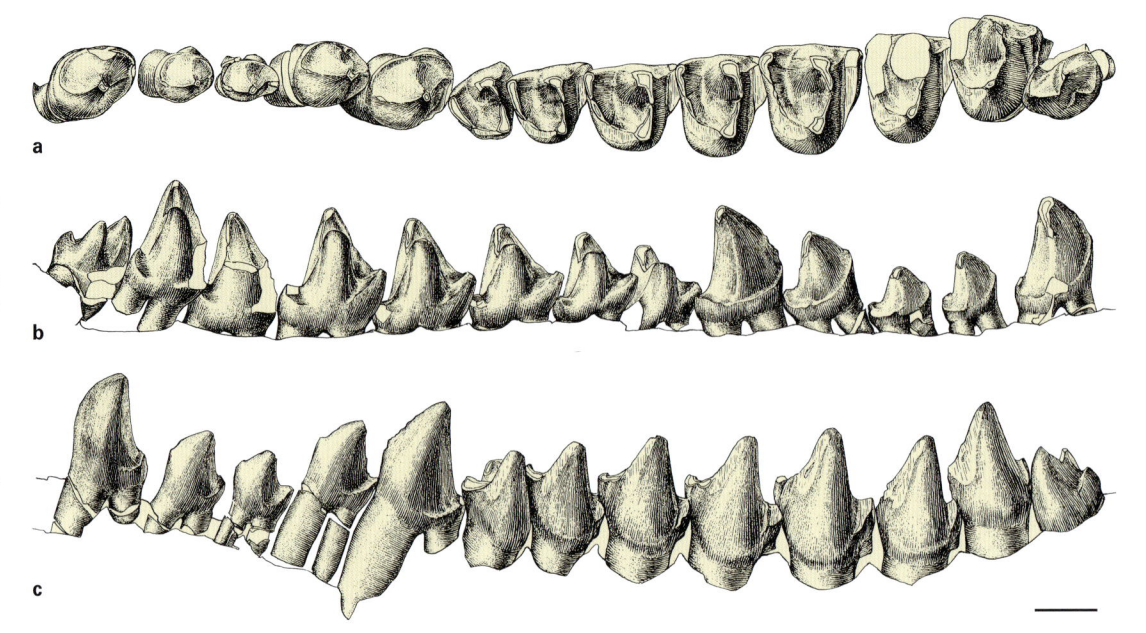

modern placental mammals, including humans, only have a maximum of three molars in each side of the jaws, eight or nine of these teeth are present in the Dryolestidae. This raises the question of the kind of food that such a dentition is adapted for. The sharp teeth of the dryolestids seem to be little suited for grinding or crushing hard plant food. The pointed cusps with their sharp cutting edges are rather useful for puncturing and disintegrating of insects, as it is seen in the Recent shrews or hedgehogs. The power produced by the jaw muscles is concentrated on the points of the tooth cusps, resulting in a very high pressure there. Thus, the cusps can puncture the hard chitinous shell of insects, like the serrations of a tomato knife puncture the smooth skin of a tomato before the actual cutting process starts. If the molars of *Dryolestes* are examined under the microscope, very fine traces of wear are found, which result from the grinding of the teeth on each other during chewing. These wear traces are not vertical, but slightly oblique, and thus indicate that the dryolestids moved their jaws also slightly transversely during chewing. Therefore, a crushing and shearing chewing process is present, which distinguishes mammals from reptiles, in which the jaw movements are usually purely vertical (orthal). The reason for this is that reptiles usually do not chew their prey, but only secure it with their pointed teeth, before they swallow it in one piece.

In contrast to the lower jaw of reptiles, which consists of a high number of separate bones, that of mammals is only made up of a single bone, the dentary. The reduction of the other jaw bones is well documented in the lineage that leads from the mammal-like reptiles to the mammals. Looking at the lower jaw of *Dryolestes* and other dryolestids, it seems to consist only of the dentary at the first glance. On closer investigation, however, a small triangular area is found in the region where the coronoid process originates (Fig. 16.4). In this area, the bone surface is slightly depressed and roughened, as it is typical for areas of contact between two bony elements. A small splint of bone was present here, which represents a rudimentary "reptilian" jaw bone, the coronoid (KREBS 1969). Since this small remnant of the coronoid was only connected to the dentary by connective tissue, it was detached and finally lost in all preserved dentaries during the decay of the soft tissues.

A small groove is visible on the inner side of the lower jaw (Fig. 16.3), which extends from the pterygoid fossa at the posterior end of the jaw to the area in which the two rami of the mandible meet, the symphysis. This is a structure that contained the remnants of the original cartilaginous rod of the lower jaw, the MECKEL´s cartilage. Apart from this cartilage, the groove probably also contained blood vessels. In the posterior part of this groove, two fine, parallel depressions are visible above and below it, which converge anteriorly (Fig. 16.4). They have been interpreted as impressions of the splenial, another one of the reptilian lower jaw bones (KREBS 1971). This wedge shaped splenial covered the deeper posterior part of the MECKEL's groove medially.

In comparison with the lower jaws of other Mesozoic mammals, the mandible of dryolestids is significantly higher. In one specimen (Fig. 16.4), the inner wall of the lower jaw is broken off, thus opening a view into the interior of the dentary and on the roots of the teeth. In relation to the height of the crowns of the teeth, the roots are gigantic, and they fill almost the entire height of the mandible. The reasons for such overly large tooth roots may be very high transversal forces during chewing, which required an especially strong implantation of the teeth.

At the posterior end of the lower jaw, the high coronoid process is found, the lateral surfaces of which served as insertion areas for powerful jaw muscles. The articular process (Processus articularis) is situated at the posterior end of the coronoid process, slightly below the highest point of the latter. Its roughly cylindrical articular head, which is oriented perpendicular to the main axis of the mandible, articulated with a deep groove in the skull in the living animal, and thus made the movements of the lower jaw possible. On the postero-ventral end of the lower jaw, a small, rod-like process is found, which is straight and directed posteriorly. This angular process (Processus angularis) is a typical character of the mammals sensu stricto, which is, for example, absent in the docodont *Haldanodon*. It is well developed in all dryolestids from the Guimarota mine, which is further evidence that this group is a representative of the stem lineage of modern mammals.

Skull: Unfortunately, only fragments are preserved of the skulls of the dryolestids. A partially preserved anterior part of a skull of *Dryolestes* has been found, which provides some information on the morphology of this animal (MARTIN 1999) (Fig. 16.6). The premaxilla indicates that the tip of the snout of *Dryolestes* was rounded. Four alveoli are visible in palatal view, which contained the single-rooted incisors. In all modern mammals, the incisors are restricted to the premaxilla. However, a fragment of an upper jaw, which is most probably referable to *Dryolestes leiriensis*, shows that the posteriormost (fifth) upper incisor was placed in the maxilla. The presence of such a so-called maxillary incisor is a primitive character that is present in several different taxa of early mammals. Thus, *Dryolestes* has five incisors in the upper jaw, which is one more than in the lower jaw. Lingually to the alveoli of the incisors, three shallow pits are present; these depressions housed the tips of the lower incisors when the jaws were closed.

Fig. 16.6. Fragment of the anterior part of a skull of *Dryolestes leiriensis* in palatal view (Gui Mam 142/75). Four alveoli for the incisors are visible in the left premaxilla. In the maxilla, the broken roots of the canine, the first premolar, the two alveoli of the third premolar and the large fourth premolar (the second premolar is reduced) are present, followed by six molars. Length of the skull fragment 2.5 cm.

As it was the case in the lower jaws, a large canine with two roots follows posteriorly behind the incisors. The first two premolars are strikingly small; the second premolar is often even reduced completely. As it was the case in the lower jaw, the third and fourth premolar are significantly larger.

The crowns of the upper molars exhibit the shape of an approximately even-sided triangle (Fig. 16.7), as it is the case in the crowns of the molars in the lower jaw. However, the bases of these triangles are situated at the lateral side of the jaw, so that the molars of the upper and lower jaws form alternating triangles. During chewing, these triangles interlock and thus cut the food items into pieces when their cutting edges slide along each other. The triangle of the molars of the upper jaw are formed by the stylocone (large cusp on the lateral side of the tooth), the paracone (cusp at the lingual side of the tooth) and the metacone (cusp at the posterior side of the tooth). Thus, the triangle of the molars of the upper jaw of dryolestids is not entirely homologous to that found in the modern (tribosphenic) mammals, since they still lack a protocone. The tribosphenic upper molar is formed by proto-, para- and metacone; the stylocone is almost completely reduced in this modern type of molar.

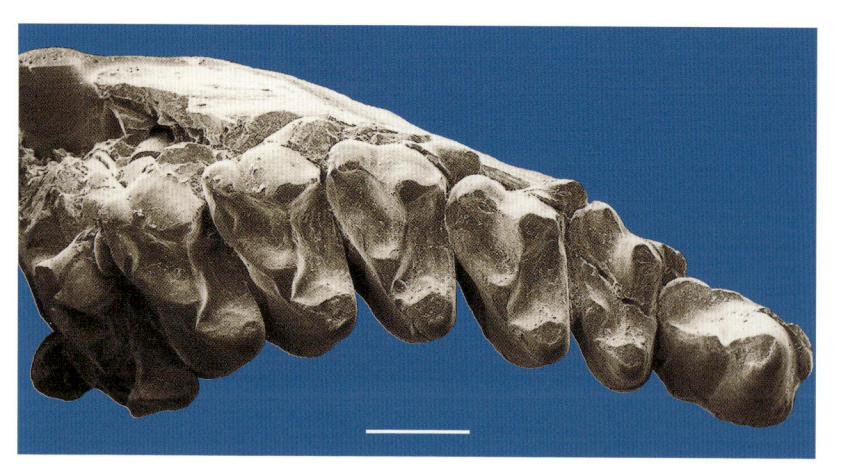

Fig. 16.7. Scanning electron-microscopic photograph of the dentition of a right upper jaw of *Dryolestes leiriensis* (Gui Mam 162/75) with the fourth premolar and six molars. *Dryolestes* is characterized by a large stylocone (the pyramid-like cusp on the upper margin of the tooth) in the molars. The tip of the jaw points to the left. Scale bar = 1 mm.

Fig. 16.8. Left upper arm bone (Humerus) of *Dryolestes leiriensis* in anterior view (Gui Mam 200). The laterally placed entepicondylar foramen and the spherical articular head of the elbow is visible at the distal end. Preserved length of the humerus 1.7 cm.

Postcranial skeleton: Some isolated limb bones are the only elements preserved of the postcranial skeleton of *Dryolestes*. An especially well preserved upper arm bone (humerus) can be referred to *Dryolestes leiriensis* on the basis of its size (Fig. 16.8). The bone exhibits a well developed deltoid crest, which indicates powerful forelimb muscles. The perpendicular articular surface with the articular facets for the ulna and radius and the entepicondylar foramen, which contained blood vessels and nerves, are well visible at the lower (distal) end.

Krebsotherium lusitanicum MARTIN 1999, the small dryolestid from the Guimarota mine

Apart from the large dryolestid, *Dryolestes leiriensis*, a second, significantly smaller dryolestid is quite common in the Guimarota mine (Figs. 16.9-16.12). The length of the lower jaw of *Krebsotherium lusitanicum* is approximately 2.5 cm, indicating a size of a large shrew for the complete animal (MARTIN 1999). The cusps of the teeth are more pointed in *Krebsotherium* than in *Dryolestes*, and the paraconid (the anteriorly directed cusp) in the molars of the lower jaw shows a conspicuous kink (Fig. 16.10). The molars of the upper jaw also exhibit significant morphological differences: the stylocone is tiny, and a ridge connecting the stylocone with the metacone is missing (Fig. 16.12). The most striking difference, however, is the difference in size; the morphological differences in the structure of the teeth is mainly of interest to the specialist.

An investigation of the microstructure of the enamel of a lower molar of *Krebsotherium* resulted in the discovery of the earliest enamel prisms known in mammals (Fig. 16.11). Whereas the needle-like crystals of the mineral apatite, of which the enamel of the teeth of vertebrates consists, forms a homogeneous matrix in most reptiles and many Mesozoic mammals, they are partially bundled in prisms in eupantotheres. The enamel prisms are formed by enamel-forming cells (ameloblasts). In the Late Jurassic eupantotheres, the prisms run parallel to each other in a straight line from the enamel dentine junction towards the outer surface of the tooth, while they form a complex, three-dimensional network in modern mammals. This network greatly strengthens the enamel of the tooth and protects it from breakage. This structurally highly developed tooth enamel made the construction of extremely resilient teeth (e.g. the gnawing teeth of rodents) possible, which allowed a very effective use of the food. Thus, the

Fig. 16.9. Right lower jaw of *Krebsotherium lusitanicum* on a matrix of coal (Gui Mam 119/74). The jaw is exposed in lingual view, so that the groove for the rudimentary MECKEL's cartilage is well visible, which runs from the pterygoid fossa (right) to the tip of the jaw (left). Length of the jaw 2.3 cm.

"invention" of enamel prisms in the Late Jurassic was an important step towards the evolutionary success of mammals.

Unfortunately, fragments of the upper jaws are the only remains preserved of the gracile skull of *Krebsotherium*, so that almost nothing can be said about its morphology. As in *Dryolestes*, no dentitions of the upper and lower jaws have been found in association in *Krebsotherium*. The association of teeth of the upper jaw with teeth of the lower jaw is an old problem in Mesozoic mammals. In these primitive mammals, the occlusion of the jaws was not as precise as in modern mammals; the teeth of the upper and lower jaws were not interlocking exactly enough to ascertain their referral to a single taxon on the basis of their corresponding morphology alone. Thus, a separate taxonomy for the upper and lower dentitions existed in dryolestids so far. However, since *Dryolestes leiriensis* and *Krebsotherium lusitanicum* are clearly distinguished by their size, morphology and relative abundance (*Dryolestes* being more common), the isolated upper and lower jaws can be referred to one of the two taxa with certainty.

Guimarotodus inflatus MARTIN 1999, a dryolestid with a powerful dentition

The robust species *Guimarotodus inflatus* is only represented by a few lower jaw fragments so far. The species name "*inflatus*" refers to the massive, inflated appearing metaconid of the molars of the lower jaw (Fig. 16.13). In contrast to *Dryolestes* and *Krebsotherium*, the metaconid is not pointed, but elongate and chisel-like. The other cusps of the premolars and molars are also very massive and robust (MARTIN 1999). This indicates that *Guimarotodus* probably fed on relatively hard food, such as heavily armored insects or shelled mollusks (e.g. snails).

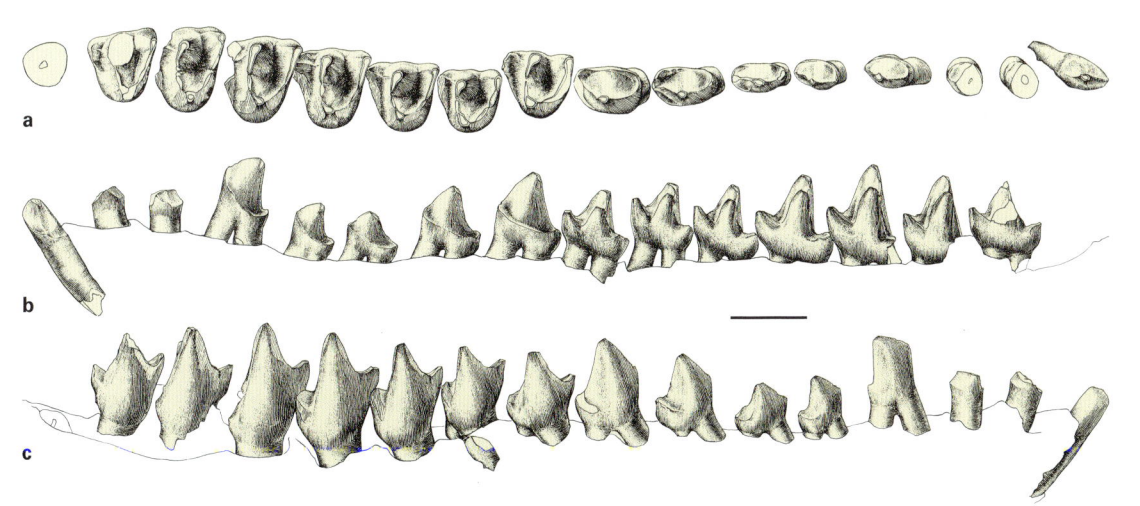

Fig. 16.10. Tooth row of a right lower jaw of *Krebsotherium lusitanicum* (Gui Mam 9/75, holotype). **a** in dorsal (occlusal) view, **b** in inner (lingual) view, **c** in outer (labial) view. In **b**, three incisors (I_{2-4}), the large canine, four premolars (P_{1-4}), and seven molars (M_{1-7}) are visible from left to right; only a fragment of the root is preserved of the rudimentary eighth molar. The teeth are principally similar to those of *Dryolestes leiriensis*, but show some differences in details (e.g. the more strongly pointed cusps). Scale bar = 1 mm. From MARTIN (1999).

Fig. 16.11. Tooth enamel of a lower molar of *Krebsotherium lusitanicum*. The outer surface of the enamel is to the right, the enamel dentine junction to the left. Scale bar = 10 μm. The enamel is formed by microscopically small, needle-like crystals of hydroxyl-apatite. In most reptiles and primitive mammals, they form a homogeneous matrix, as can be seen in the extreme right of the picture. In the dryolestids from the Guimarota mine, they are partially bundled into fibrous units, the enamel prisms. From MARTIN (1999).

The best preserved jaw of *Guimarotodus inflatus* exhibits an unusual character. Behind the canine, not four, but five premolars are present: an additional tooth is found very closely behind the second regular premolar, which is similar to the latter in its size and morphology. However, on closer inspection, this tooth proves to be heavily worn. Since the degree of wear is correlated with the time a tooth has been in use, this tooth must have been in the jaws much longer than the other premolars. Morphological differences are also evident on closer inspection. The tooth corresponds to a milk molar in terms of its morphology (see "tooth replacement"), and not to a premolar of the permanent dentition. Thus, this tooth obviously represents a milk tooth that has not been shed, but remained in the jaw together with its successor of the permanent dentition (a phenomenon that is sometimes found in mammals and also humans). This explains the heavy wear of the tooth, since the milk tooth spent a much longer time in the jaw than the premolars of the permanent dentition.

Fig. 16.12. Scanning electron-microscopic photograph of the dentition of a left upper jaw of *Krebsotherium lusitanicum* (Gui Mam 73/79) with the fourth premolar and eight molars (the eight molar, in the picture to the right, is rudimentary). In contrast to *Dryolestes*, the stylocone is strikingly small in *Krebsotherium* (the small cusp in the middle of the upper margin of the molars). The tip of the jaw points to the left. Scale bar = 1 mm.

Tooth replacement in dryolestids

Seven lower jaws and two upper jaws of juveniles are present in the material from the Guimarota mine, which have complete or partial milk dentitions preserved (Figs. 16.14 and 16.15). The jaw bone is not completely mineralized in these jaws, which is evident from the cancellous appearance of the bone surface. Milk dentitions of Mesozoic mammals are extremely rare, and they can provide important information on the evolution of mammalian dentitions in general. All placental mammals, including humans, have a diphyodont tooth replacement, i.e. the teeth are only replaced once during a lifetime. However, this is only the case in the teeth in front of the molars: incisors, canine, and premolars. The molars themselves belong to the first generation of teeth and are not being replaced at all.

In contrast to mammals, reptiles replace their teeth repeatedly throughout their lifetime in reoccurring waves. This is only possible because their teeth are usually of a simple conical shape and isodont, i.e. they all have a very similar morphology (e.g. in a crocodile). Such a permanent tooth replacement would be impossible in mammals, given the complex structure of the molars and their precise occlusion, since such a dentition would be useless with gaps in it. Therefore, the diphyodont tooth replacement of modern placentals must have evolved from the permanent tooth replacement of reptiles during the evolutionary history of their ancestors. Only fossil mammals can provide us with evidence and insights into when and how this transformation took place.

In the juvenile jaws, which exhibit different stages of tooth replacement, it is evident that the tooth replacement was also restricted to the teeth in front of the molars in dryolestids (Fig. 16.15). In the juveniles, only one or two true molars are developed; this number is then increased to eight or nine during growing up. The sequence of tooth replacement still reflects the alternating tooth replacement pattern found in reptiles: alternately, teeth are replaced earlier or later (MARTIN 1997).

Apart from its evolutionary significance, the tooth replacement pattern of dryolestids also has some indirect implications for their reproduction. Paired marsupial bones have been found in the pelvic region of the skeleton of *Henkelotherium* (HENKEL & KREBS 1977, KREBS 1991). This has repeatedly led to the assumption that eupantotheres had a pouch, which was used for raising the young, as it is the case in marsupials. However, marsupial bones are not necessarily correlated with the presence of a pouch, since they are

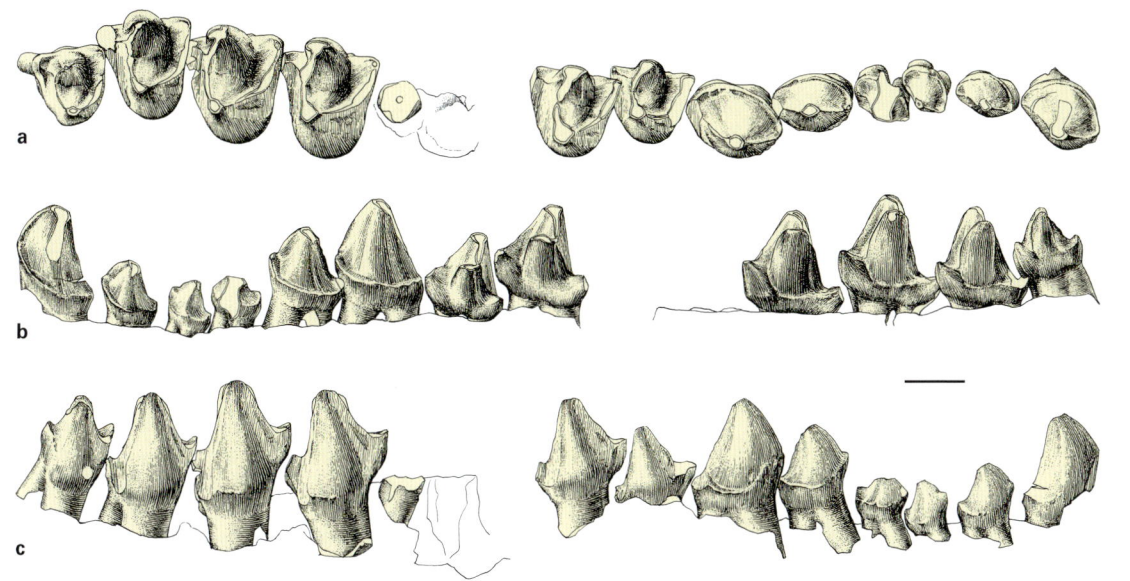

Fig. 16.13. Tooth row of a right lower jaw of *Guimarotodus inflatus* (Gui Mam 121/75, holotype). **a** in dorsal (occlusal) view, **b** in inner (lingual) view, **c** in outer (labial) view. In **b**, the canine, the first and second premolar, the milk tooth of the second premolar, the third and fourth premolar, first and second molar, root fragments of the fourth molar, and the fifth to eighth molar are shown from left to right. The teeth of *Guimarotodus* appear to be especially robust and strong. Scale bar = 1 mm. From MARTIN (1999).

imbedded in the connective tissue of the abdominal wall in the living animal, and are thus morphologically not directly related to the pouch. They rather represent a primitive character that is found in many basal mammals, and not only in marsupials. On the other hand, the juveniles of marsupials are characterized by a very unusual tooth replacement pattern: only the 3rd premolar is replaced in marsupials, and not all teeth in front of the molars; the other teeth are either persistent milk teeth, or teeth of the permanent dentition that were not preceded by milk teeth. This strongly modified tooth replacement pattern is known since the Late Cretaceous in the fossil record (CIFELLI et al. 1996). It is connected with the behaviour of the juveniles in the pouch of the mother: the juveniles of marsupials, which are only a few millimeters large, attach themselves very firmly to the teats of the mother, which then fills their entire mouth, for several weeks. Because of the pressure of the teat on the epithelium of the mouth, the tooth germs are formed belated, and thus, the tooth replacement is incomplete. Since the replacement of the antemolar teeth in dryolestids, however, follows the normal pattern, a marsupial-like reproduction with a firm attachment of the young to the teats of the mother seems to be rather unlikely for this group (MARTIN 1997, 1999).

The lifestyle of the dryolestids

Although no direct evidence for the presence of fur in dryolestids has been found, due to the circumstances of preservation in the Guimarota mine, it can safely been assumed that they were furry, warm-blooded mammals. Such a highly differentiated dentition, as it is found in the dryolestids, which is designed for chewing, does only make sense if the food had to be thoroughly processed for fuelling a warm-blooded metabolism. Furthermore, impressions of fur are preserved in mammals from the Early Cretaceous of Liaoning (north-eastern China) (HU et al. 1997). The dryolestids can be envisioned as small, insectivorous (insect-eating) mammals of a head-and-

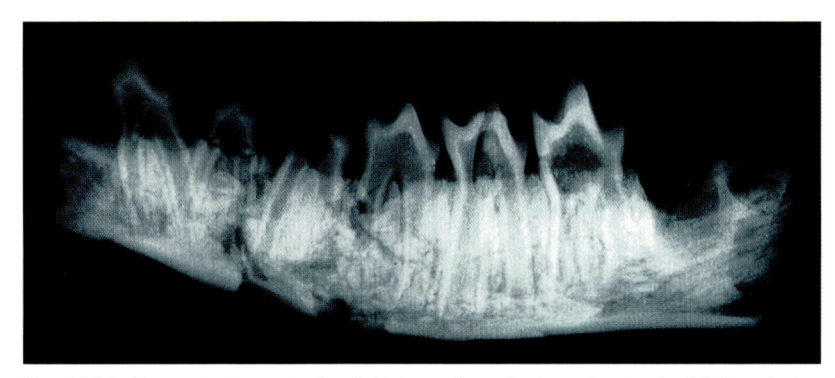

Fig. 16.14. X-ray photograph of a right lower jaw of a juvenile dryolestid *(Dryolestes leiriensis)*, with the milk dentition. The strongly separated roots of the milk teeth are well visible in the X-ray photograph. Only one molar was fully developed in this juvenile; behind the molar, the alveoli for the germ of the second molar can be seen. Length of the jaw: 1.1 cm. X-ray photograph: J. HABERSETZER, Forschungsinstitut Senckenberg (Frankfurt am Main).

Fig. 16.15. Left lower jaw of a juvenile dryolestid (*Dryolestes leiriensis*, Gui Mam 112/76), showing tooth replacement. Four incisors can be seen from left two right; the first, third and fourth of these are milk teeth. The germ of a permanent replacement tooth is visible close to the root of the fourth milk-incisor. Behind this tooth, the milk canine with its strongly separated roots is present. The erupting tooth is the first premolar; it is followed by the second milk-premolar with separated roots. The crown of the following third milk-premolar later has been removed in the lab, revealing the tip of the permanent third premolar between its remaining roots. The posteriormost teeth are the fourth milk-premolar and the first true molar; behind that, the jaw is broken. Scale bar = 1 mm.

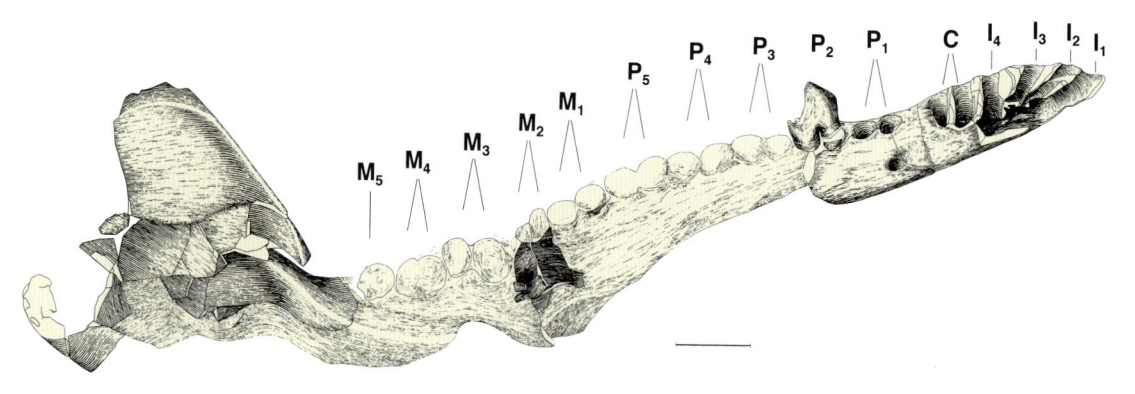

Fig. 16.16. Right lower jaw of the primitive "peramurid" from the Guimarota mine. The areas shown in thin lines are only preserved as plastic casts of the mold of the bone in the coal matrix. With the exception of the second premolar (P_2), all teeth are missing. However, the alveoli for the four single-rooted incisors, and the double alveoli for the canine and the P_1 are well visible in front of the preserved tooth. Behind it, the double alveoli for the P_3-P_5 and the first to fourth molar (M) are present; M_5, the last molar, only has a single alveolus. Scale bar = 1 mm.

body length of 10-25 cm, which were roughly shrew-like. We know that the closely related *Henkelotherium* had high, pointed claws, which after KREBS (1991) indicate that it was a good climber. However, small mammals are all more or less well able to climb; for them, it does not make much of a difference whether they move on the ground or on the rough bark of a tree. Therefore, this does not necessarily represent a strict adaptation for an arboricolous (tree-climbing) lifestyle. Thus, the dryolestids were probably nocturnal, as it is typical for many small mammals, and lived in the shrubbery, where they hunted small invertebrates, such as insects, larvae and snails, but also small vertebrates. The story that the small Mesozoic mammals endangered the gigantic dinosaurs by eating their eggs during the night, must be regarded as a fairy tale. The small mammals would have hardly been able to crack open the shells of dinosaur eggs, which were enormous in relation to them.

The primitive "peramurid"

W. G. KÜHNE (1968) already noted very small lower molars with especially pointed cusps and a relatively large talonid in the material from the first excavation campaign, in the end 1950ies and early 1960ies. He referred an especially slender and gracile fragmentary lower jaw, which had lost all teeth with the exception of the second premolar to these molars (Fig. 16.16). KÜHNE (1968) tentatively placed the teeth and the lower jaw in the genus *Peramus*, but did not analyze the material in detail. This referral is noteworthy, since *Peramus*, a eupantothere from the latest Jurassic of southern England, exhibits an initial development of a talonid basin in the molars of its lower jaw. Thus, *Peramus* is more closely related to the lineage that leads towards modern mammals than either the Dryolestidae or the Paurodontidae.

Since only three or four fragmentary teeth and the jaw fragment of this enigmatic animal were present, we examined the approximately 10,000 isolated teeth resulting from the screenwashing of the coals of Guimarota, in the hope to find further teeth of this taxon. And really, 48 upper and lower molars of the primitive "peramurid" were found among the thousands of mammalian and reptilian teeth (Figs. 16.17 and 16.18). It was a special piece of luck that furthermore an anterior fragment of a lower jaw with three premolars was found among the so far undetermined jaw remains. The tooth formula of this new mammal can now be reconstructed with the help of the new fragment and the jaw already described by W. G. KÜHNE. It had four incisors, one canine, five premolars, and five molars (MARTIN in press). This tooth formula is significantly different from that of the dryolestids and paurodontids: the number of molars is less, but one additional premolar is present in comparison with the Dryolestidae, which represents the primitive condition for modern mammals, according to MCKENNA (1975). The molars in the new material confirms that the primitive "peramurid" from the Guimarota mine had a slightly elongated talonid with a large main cusp. A small additional cusp is found on the ridge that extends from the metaconid to this main cusp (Fig. 16.17); a character that is also present in *Peramus*. In general, however, the talonid of the primitve "peramurid" lacks any trace of an incipient basin and therefore is more primitive than that of *Peramus*. This is not unexpected, since the mammals from Guimarota are somewhat older geologically. Thus, the new taxon can be regarded as a primitive stem-lineage representative of Zatheria MCKENNA 1975, the group that also comprises the modern tribosphenic mammals.

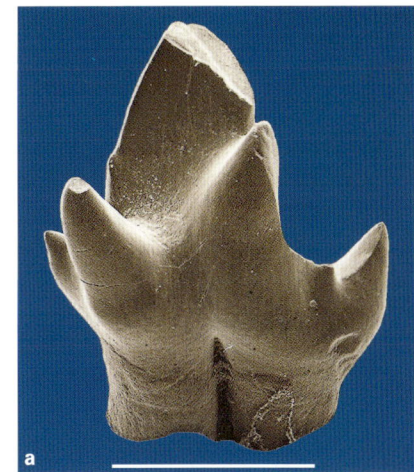

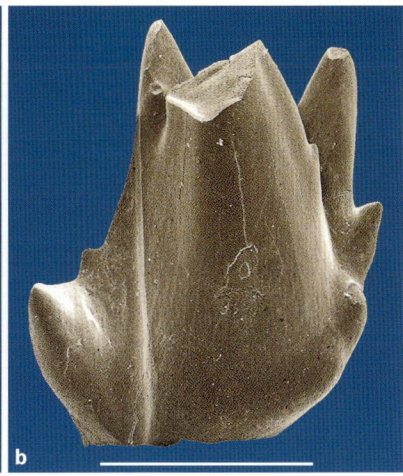

Fig. 16.17. Scanning electron-microscopic photograph of a right lower molar of the primitive "peramurid" (Gui Mam 1002); **a** in lingual view, **b** in labial view. In contrast to the dryolestids, the talonid (on the right side in a, on the left in b) is elongated and has an additional small cusp. Scale bar = 0.5 mm.

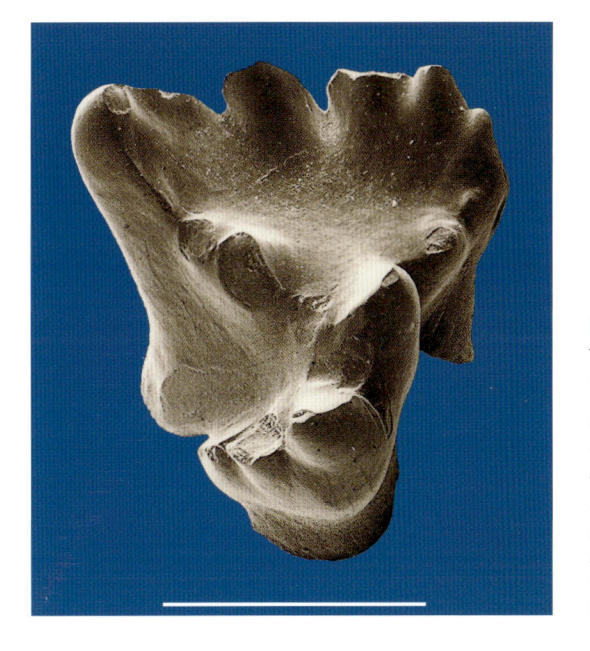

Fig. 16.18. Scanning electron-microscopic photograph of a right upper molar of the primitive "peramurid" (Gui Mam 1033), in mesio-occlusal view. The upper molars are characterized by a small, pointed stylocone and a striking elongation of the marginal cusps. Scale bar = 0.5 mm.

References

BONAPARTE, J. F. (1986): Sobre *Mesungulatum houssayi* y nuevos mammíferos Cretacicos de Patagonia, Argentina. – Actas IV Congreso Argentino de Paleontología y Bioestratigrafía **2**: 48-61; Mendoza.
– (1990): New Late Cretaceous mammals from the Los Alamitos Formation, Northern Patagonia. – National Geographic Research **6**: 63-93.
– (1994): Approach to the significance of the Late Cretaceous mammals of South America. – Berliner geowissenschaftliche Abhandlungen E **13**: 31-44.
CIFELLI, R. L.; ROWE, T. B.; LUCKETT, W. P.; BANTA, J.; REUBEN, R. & HOWES, R. I. (1996): Fossil evidence for the origin of the marsupial pattern of tooth replacement. – Nature **379**: 715-718.
CLEMENS, W. A. & LEES, P. M. (1971): A review of English Early Cretaceous mammals. – Zoological Journal of the Linnean Society **50**, suppl. 1: Early Mammals: 117-130.
FREEMAN, E. F. (1976): Mammal teeth from the Forest Marble (Middle Jurassic) of Oxfordshire, England. – Science **194**: 1053-1055.
HENKEL, S. & KREBS, B. (1969): Zwei Säugetier-Unterkiefer aus der Unteren Kreide von Uña (Prov. Cuenca, Spanien). – Neues Jahrbuch für Geologie und Paläontologie, Monatshefte **1969**: 449-463.
HENKEL, S. & KREBS, B. (1977): Der erste Fund eines Säugetier-Skelettes aus der Jura-Zeit. – Umschau Wissenschaft und Technik **77**: 217-218.
HU, Y.-M., WANG, Y.-Q., LUO, X. & LI, C.-K. (1997): A new symmetrodont mammal from China and its implications for mammalian evolution. – Nature **390**: 137-142.
KREBS, B. (1969): Nachweis eines rudimentären Coronoids im Unterkiefer der Pantotheria (Mammalia). – Paläontologische Zeitschrift **43**: 57-63.
– (1971): Evolution of the mandible and lower dentition in dryolestids (Pantotheria, Mammalia). – Zoological Journal of the Linnean Society **50**, suppl. 1: Early mammals: 89-102.
– (1987): The skeleton of a Jurassic eupantothere and the arboreal origin of modern mammals. – Fourth Symposium on Mesozoic terrestrial Ecosystems, short papers: 132-137; Drumheller (Tyrrell Museum).
– (1991): Das Skelett von *Henkelotherium guimarotae* gen. et sp. nov. aus dem Oberen Jura von Portugal. – Berliner geowissenschaftliche Abhandlungen A **133**: 1-110.
KÜHNE, W. G. (1968): Kimmeridge mammals and their bearing on the phylogeny of the Mammalia. – In: DRAKE, E. T. [ed.] Evolution and Environment: 109-123, New Haven, London (Yale University Press).
LILLEGRAVEN, J. A. & MCKENNA, M. C. (1986). Fossil mammals from the "Mesaverde" Formation (Late Cretaceous, Judithian) of the Bighorn and Wind River

Basins, Wyoming, with definitions of Late Cretaceous North American land-mammal "ages". – American Museum Novitates **2840**: 1-68.

MARSH, O. C. (1878): Fossil mammal from the Jurassic of the Rocky Mountains. – American Journal of Science 3, **XV**: 459.

– (1879): Notice of new Jurassic mammals. – American Journal of Science 3, **XVIII**: 396-398.

MARTIN, T. (1995): Dryolestidae from the Kimmeridge of the Guimarota coal mine (Portugal) and their implications for dryolestid systematics and phylogeny. – In: SUN, A. & WANG, Y. [eds.] Sixth Symposium on Mesozoic Terrestrial Ecosystems and Biota, short papers: 229-231, Beijing (China Ocean Press).

– (1997): Tooth replacement in Late Jurassic Dryolestidae (Eupantotheria, Mammalia). – Journal of Mammalian Evolution **4**: 1-18.

– (1999): Dryolestidae (Dryolestida, Mammalia) aus dem Oberen Jura von Portugal. – Abhandlungen der Senckenbergischen naturforschenden Gesellschaft **550**: 1-119.

– (in press): New stem-lineage representatives of Zatheria (Mammalia) from the Late Jurassic of Portugal. – Journal of vertebrate Paleontology.

MCKENNA, M. C. (1975): Toward a phylogenetic classification of the Mammalia. – In: LUCKETT, W. P. & SZALAY, F. S. [eds.] Phylogeny of the Primates: 21-46, New York (Plenum Press).

PROTHERO, D. R. (1981): New Jurassic mammals from Como Bluff, Wyoming, and the interrelationships of non-tribosphenic Theria. – Bulletin of the American Museum of Natural History **167**: 277-326.

SCHUDACK, M. E. (1993): Charophyten aus dem Kimmeridgium der Kohlengrube Guimarota (Portugal). Mit einer eingehenden Diskussion zur Datierung der Fundstelle. – Berliner geowissenschaftliche Abhandlungen E **9**: 211-231.

– (1995): Neue mikropaläontologische Beiträge (Ostracoda, Charophyta) zum Morrison-Ökosystem (Oberjura des Western Interior, USA). – Berliner geowissenschaftliche Abhandlungen E **16**: 389-407.

– (1996): Ostracode and charophyte biogeography in the continental Upper Jurassic of Europe and North America as influenced by plate tectonics and paleoclimate. – In: MORALES, M. [ed.] The Continental Jurassic. – Museum of Northern Arizona Bulletin **60**: 333-341.

SIMPSON, G. G. (1928): A catalogue of the Mesozoic Mammalia in the Geological Department of the British Museum. – 215 pp., London (British Museum [Natural History]).

– (1929): American Mesozoic Mammalia. – Memoirs of the Peabody Museum of Yale University **3**: 1-235.

THENIUS, E. (1989): Zähne und Gebiss der Säugetiere. – Handbuch der Zoologie, NIETHAMMER, J., SCHLIEMANN, H. & STARCK, D. [eds.] **8, 56**: 513 pp., Berlin, New York (Walter de Gruyter).

17 The henkelotheriids from the Guimarota mine

BERNARD KREBS

The henkelotheriids, the second group of eupantotheres found in the coals of the Guimarota mine, are less abundant than the dryolestids. They are characterized by a dentition with three or four premolars and five to six molars in the upper and lower jaws (sometimes, a seventh molar is present in the lower jaw; if present, this tooth is rudimentary). In contrast to the extremely shortened molars of the dryolestids, those of the henkelotheriids are approximately as long as they are wide. Furthermore, the roots of the lower molars of henkelotheriids are of approximately equal size, whereas the anterior one is very massive, and the posterior root is slender in the dryolestids. Further distinctive characters are present in the morphology of the teeth of the henkelotheriids, but these characters will not be discussed in detail here.

The Henkelotheriidae are represented by two genera with one species each in the Guimarota mine: *Henkelotherium guimarotae* and *Drescheratherium acutum*. The name *Henkelotherium* was given in memoriam of SIEGFRIED HENKEL, who organized and led the excavations in the Guimarota mine (see chapter on the history of the excavations). The name *Drescheratherium* acknowledges the work of the preparator ELLEN DRESCHER, who skillfully prepared and preserved many of the fossils from the Guimarota mine (see chapter on the preparation).

Henkelotherium guimarotae

This taxon is of extraordinary importance, since it is based on an almost complete skeleton. It is the most remarkable find from the Guimarota mine, being the first discovery of an articulated skeleton of a mammal from the Jurassic. It is furthermore the geologically oldest skeleton of a representative of the Theria, the group of mammals, which also includes the modern mammals, the marsupials and placental mammals and their direct ancestors. The skeleton of *Henkelotherium* documents the level of organization of the Theria 150 million years ago and shows, which adaptations were present at the basis of the radiation of modern mammals (KREBS 1991).

The fossil was found on December 15th, 1976, by the Portugese helper GRAZIELA. At this time, none of the scientists from Berlin was in Portugal. The forewoman, Dona ENCARNAÇÃO, recognized that more than a jaw was present. She assumed that it was a skull. Only the preparation, which started in the beginning of 1977, showed that a whole skeleton was present. The very laborious preparation and conservation of the find is described in the chapter on the preparation.

The skeleton of *Henkelotherium* is now embedded in an artificial matrix (Figs. 17.1 and 17.2). The animal was very small, approximately the size of a mouse. The outline of the artificial matrix reflects the shape of the slab of coal that originally contained the specimen. It was a lucky circumstance that almost the entire skeleton was present on this small, randomly shaped area. However, a part of the skull lay outside the slab and was thus not preserved. A smaller piece of coal, which came apart from the main slab while splitting the coal, contained two fingers of the right hand and some skull bones; these elements were also transferred onto the artificial matrix.

The skeleton is largely preserved in articulation and mainly exhibits the ventral side. The bones of the cervical vertebral column behind the skull and the anterior trunk elements have been slightly disarticulated, probably by the eruption of decay gases. The right shoulder girdle and the right forelimb are visible, with the upper arm, lower arm and a part of the hand. The left forelimb is unfortunately missing. The last pectoral vertebrae, the lumbar vertebrae and the sacrals are still in articulation. Then, the vertebral column is slightly disarticulated; the sacral vertebrae have lost their connection with the pelvis. The tail, which is recurved in an extended arc, begins in the area of the pelvis. Both hindlimbs are splayed outwards. While the left foot is missing, the right one was fortunately flexed in a way that allowed its almost complete preservation on the slab.

Fig. 17.1. *Henkelotherium* (Gui Mam 138/76, holotype). The prepared skeleton. The specimen is only 5.5 cm high. The skeleton is mainly exposed in ventral view (see also Fig. 17.2).

Skull (Fig. 17.3): The only identifiable elements of the skull are the marginal parts of the right side. The most striking element is the right ramus of the mandible, which tipped over onto the palate, so that the right upper jaw dentition is exposed along the ventral margin of the lower jaw (Fig. 17.3). Apart from the articular process, the lower jaw exhibits a high coronoid process and a strongly developed angular process posteriorly, similar to the situation in Recent insect eaters. However, several primitive characters are also found on the exposed medial side: a groove that extends over the complete length of the lower jaw, which contained MECKEL's cartilage (the cartilage that originally forms the lower jaw in early ontogenesis; only present in the embryos of modern mammals). Behind the last molar, a small, slightly depressed triangular area is found; this is the contact area for a rudimentary coronoid, a remaining reptilian lower jaw bone (KREBS 1971). However, the elements that form the primary jaw joint in reptiles are no longer visible; they were obviously already incorporated into the middle ear as sound-conducting bones in *Henkelotherium*.

In this context, the recent identification of the petrosal on the smaller piece of coal mentioned above, is noteworthy (Fig. 17.3). The petrosal encloses a spiraled cochlea. This is the oldest record of a spiraled cochlea in mammals. This feature obviously represents a typical character of the Theria, since the cochlea is straight or only slightly curved in all other mammals, including the monotremes (KREBS & HU in press).

In Fig. 17.3, one bony element is striking, which is disarticulated from the rest of the skull and lies below the jaws. This is the right jugal, which here formed a significant part of the jugal arch.

Dentition (Fig. 17.4): Because of the poor preservation of the skull, only the dentition of the right upper jaw is present, starting with the last incisor. Only two posteriormost premolars and five molars are preserved in their original position. A sixth molar was probably present, but it is not preserved. The teeth of the right lower jaw are preserved in situ, from the canine to the last (seventh) molar, but the canine and the two anterior premolars are only visible on the backside of the slab (Fig. 17.4). The tooth formula of *Henkelotherium* is: probably five upper and four lower incisors, one canine and four premolars in both the upper and lower jaw, and five or six molars in the upper jaw, and six molars in the lower jaw (in the present skeleton, a rudimentary seventh molar is found in the lower jaw; however, only six molars are present in other mandibles of *Henkelotherium*). The general morphology of the molars is typical for eupantotheres, and their balanced proportions, their number, and the fact that the first molar is equal in size to the last premolar characterize a henkelotheriid. In comparison with the situation found in *Drescheratherium*, the upper canine is relatively small and blunt. In both the upper and lower jaw, the second premolar is tiny and even has only one root in the upper jaw, which indicates a beginning reduction of this tooth.

Such a dentition with molars that pierce the food items with their cusps and then cut them up

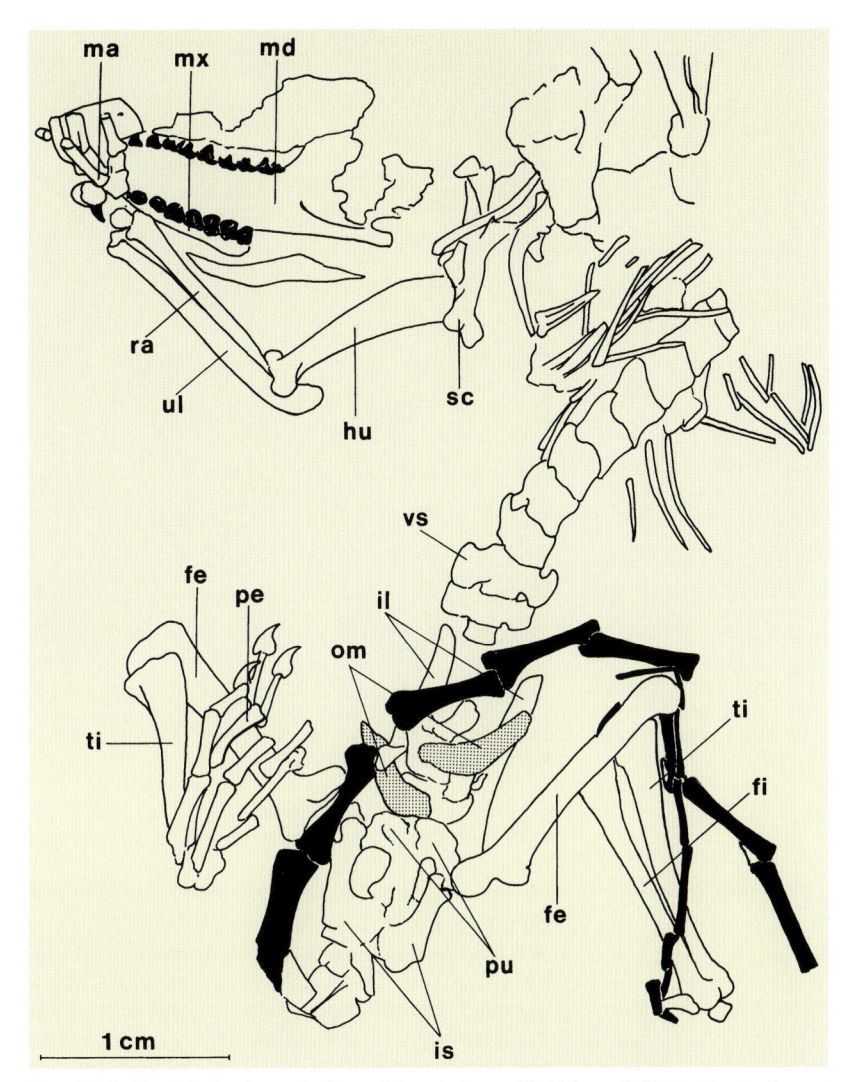

Fig. 17.2. *Henkelotherium*. Outline of the skeleton (Gui Mam 138/76). The teeth and the elongate caudal vertebrae are illustrated in black, the marsupial bones are stippled. fe femur, fi fibula, hu humerus, il ilium, is ischium, ma manus (hand), md mandible (lower jaw), mx maxilla (upper jaw), om ossa marsupialia (marsupial bones), pe pes (foot), pu pubis, ra radius, sc scapula (shoulder blade), ti tibia, ul ulna, vs vertebrae sacrales. From HENKEL & KREBS (1977).

Fig. 17.3. *Henkelotherium* (Gui Mam 138/76). Detail photograph of the skull and forelimb. Scale in mm.

with their cutting edges are well suited for a diet consisting of insects.

Shoulder girdle and forelimbs (Fig. 17.5): The shoulder girdle consists only of the scapula (shoulder blade) and the clavicle. An interclavicle and a coracoid, both typical reptilian elements that are still present in other contemporaneous mammals and monotremes (egg-laying mammals), are not present. The coracoid is reduced to a coracoid process on the shoulder blade. The flat, elongate shoulder blade exhibits an approximately medially positioned spina scapulae (a pronounced ridge), and the clavicle is developed as a thin, rod-like element. This shoulder girdle is very advanced; it is comparable with that of small Recent marsupials and placental mammals. The reduction of the ventral elements of the shoulder girdle, the procoracoid, coracoid, and interclavicle, indicates a narrow thorax and forelimbs which were drawn up towards the body.

The forelimb is also advanced: the humerus (upper arm bone) is relatively long and gracefully curved, with the articular facets for the radius and

Fig. 17.4. *Henkelotherium* (Gui Mam 138/76). Dentition. Top: dentition of the right upper jaw, with canine and first premolar (tilted sideways), the third and fourth premolar and five molars (occlusal view). Below: dentition of the right lower jaw, with third and fourth premolar and seven molars (in lingual view). Scale bar = 1 mm. From KREBS (1991).

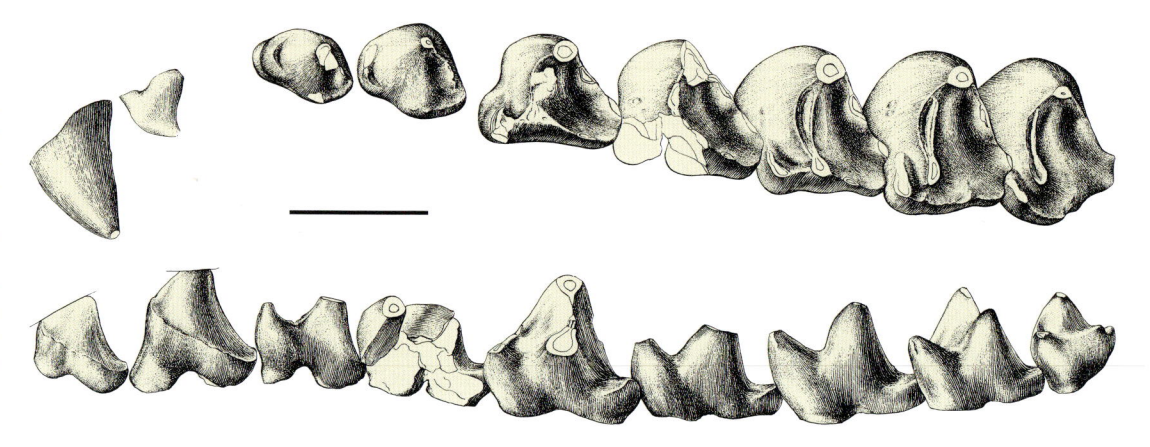

ulna being separated, and the lower arm is still slightly longer than the upper arm. The hand will be described below.

Pelvis and hindlimbs (Fig. 17.6): In the pelvis, the ilia are considerably elongated anteriorly, resulting in a significant distance between the attachment of the sacral vertebrae and the articular socket for the femur. The ischia and pubes are stocky and exhibit a large foramen obturatorium on either side. The pelvis of *Henkelotherium* is very different from that of monotremes, but it is very similar to the pelvis of modern viviparous mammals.

Two broad, slightly curved bones that are situated in front of the pelvis are noteworthy. The bone on the right side is in contact with the pubis, while that on the left side is slightly displaced anteriorly. These are the marsupial bones (Ossa marsupii, or rather epipubes). These elements were identified for the first time in an early representative of the Theria. Marsupial bones are known in Recent marsupials and monotremes. They are present in both sexes and also in several marsupials that lack a pouch, such as the didelphids. Thus, marsupial bones are not necessarily correlated with the presence of a pouch, but rather support the ventral musculature. Marsupial bones have been found in several groups of extinct mammals and even in early placental mammals. The presence of these so-called marsupial bones is thus obviously a character of all mammals, which is only reduced in advanced placentals, probably in connection with the prolonged time that the young is carried by the mother.

The femur is a long and slender bone; the ball-shaped femoral head is offset from the shaft of the bone at a distinct angle. The tibia and fibula are also long and slender. Such a hindlimb is principally not different from that of a small didelphid or an insect-eater. The offset femoral head enables the animal to move the legs underneath the body to achieve a fully erect gait.

The foot is especially interesting. The tarsus is only incompletely preserved on both sides, and the rest of the left foot is missing altogether, but the right foot is completely preserved, apart from the last phalanx of the second digit. In total, the foot is surprisingly long. Especially striking is the great length of the penultimate phalanges of the five toes. The ungual phalanges are very large, strongly curved, pointed, and compressed transversely. These features were certainly enhanced by keratinous claws. There is no evidence that the slightly shorter first toe was extendable or even opposable.

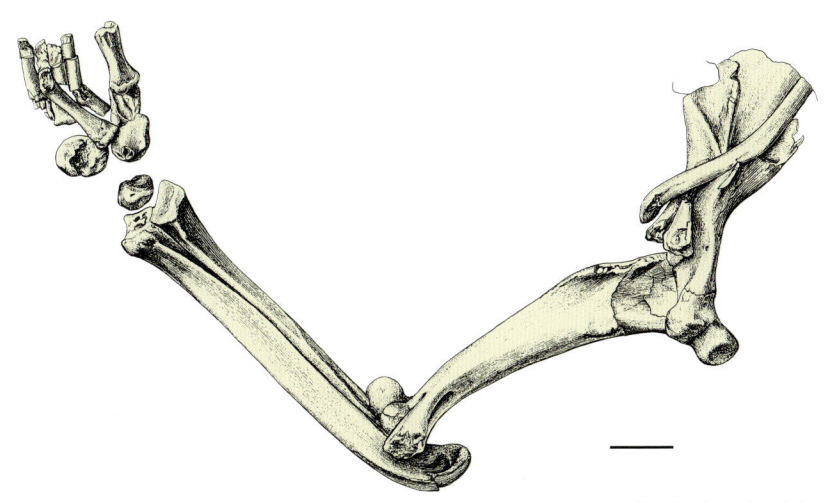

Fig. 17.5. *Henkelotherium* (Gui Mam 138/76). Shoulder girdle and forelimb of the right side. The shoulder blade (left) has a well developed crest; the clavicle is reduced to a slender rod. The upper arm and the ulna and radius are long and slender. Scale bar = 2 mm. From KREBS (1991).

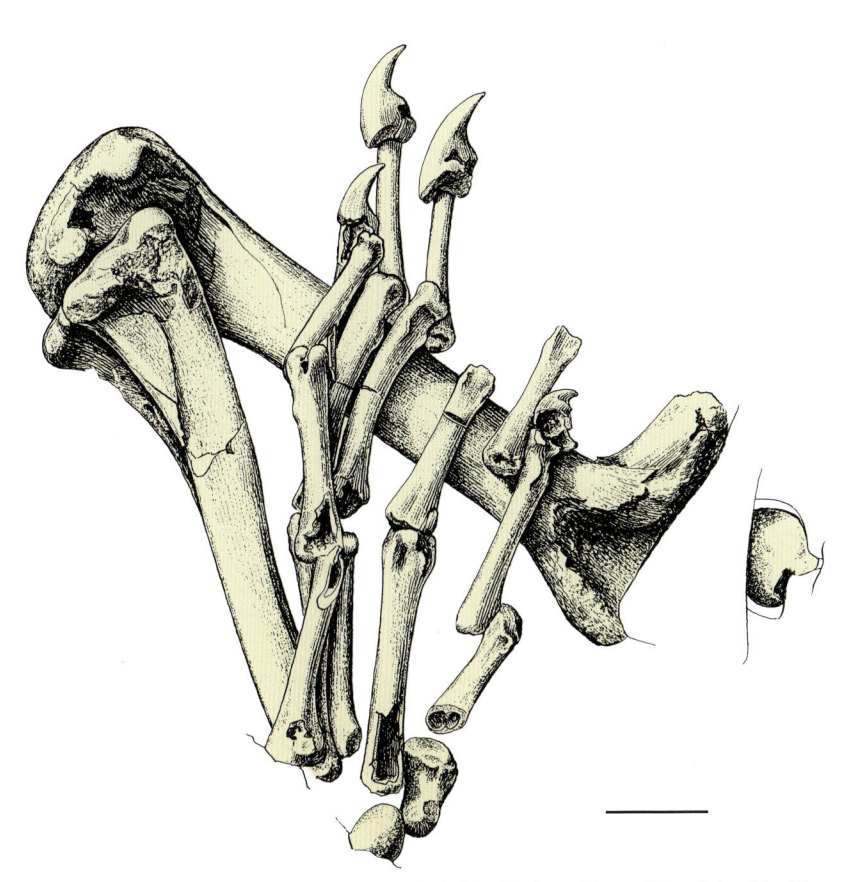

Fig. 17.6. *Henkelotherium* (Gui Mam 138/76). Hindlimb and foot of the right side. The foot is situated above the femur. Note the length of the foot, the relatively long penultimate phalanges and the ungual phalanges, which are large, significantly curved and pointed. Scale bar = 2 mm. From KREBS (1991).

Fig. 17.7. *Henkelotherium*. Life reconstruction. The mouse-sized, insectivorous, arboreal animal climbs on a branch of a ginkgo tree. Such trees are known from the Guimarota mine, but their leaves were slightly different from the ones of the recent ginkgo figured here. (Drawing by ELKE GRÖNING).

The peculiarities of the foot are also found in the hand, two fingers of which are preserved on the smaller slab of coal mentioned above. The fingers also have elongate penultimate phalanges and large, curved, pointed and narrow ungual phalanges.

This kind of hands and feet are only known from forms that climb with the help of their claws among small modern mammals. This kind of locomotion is called claw climbing, in contrast to the clutching climbing of primates, in which the animals use their opposable thumb and first toe to cling to branches and tree trunks.

Tail (Fig. 17.2): The tail usually plays an important role in climbing animals. In the present skeleton, the first six to seven short caudal vertebrae, which are preserved in the vicinity of the pelvis and are partially obscured by other bony elements, are followed by vertebrae that rapidly increase in length in the anteriorly curved portion of the tail. They finally reach a length that corresponds to the length of three dorsal vertebrae. The tail meets the margin of the slab of coal in the lower right corner; two to three vertebrae are missing here. The rest of the tail is again preserved on the slab, next to the left hindlimb. These posterior caudal vertebrae are very slender, but still extremely long. Thus, *Henkelotherium* had a very long tail, the length of which was not achieved by an increase in the number of caudals, but by an elongation of the present elements.

Such a morphology is found in modern mammals in forms that use their tail as a steering organ in jumping from tree to tree. The best known model for this is the squirrel.

In summary, *Henkelotherium* is a small, insectivorous, arboreal animal, which easily climbed tree trunks with the help of its claws, nimbly moved on small branches with its fully erect and narrow gait, and was able to carry out precise jumps with the help of its tail (see Fig. 17.7).

If we assume that the skeletal morphology of *Henkelotherium* is representative for the organizational level of the Theria in the Late Jurassic,

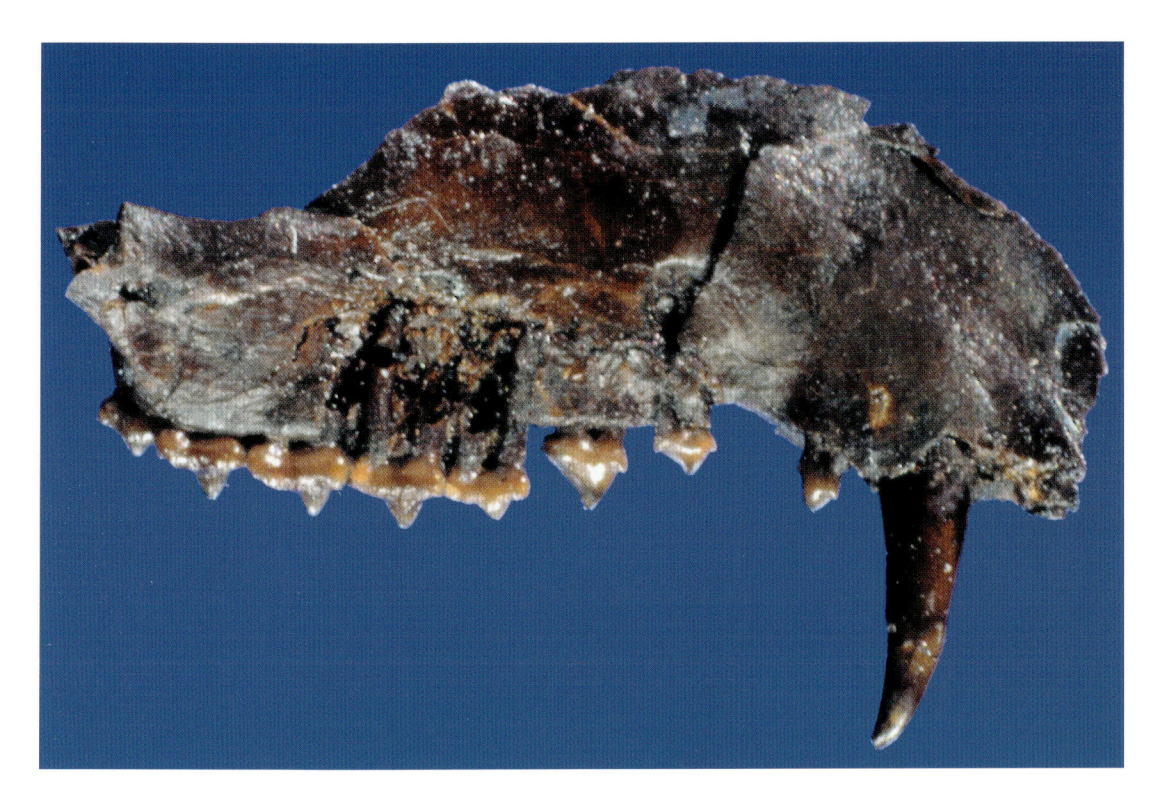

Fig. 17.8. *Drescheratherium* (Gui Mam 4/73, holotype). Right upper jaw. The extremely long, pointed canine is striking. The length of the specimen is only 12 mm.

this find provides the first paleontological evidence for the controversial idea that marsupials and placental mammals originated from arboreal animals. However, we now at least know with certainty that the characters, which are considered to be characteristic for the Theria, were already present in their ancestors 150 million years ago.

Drescheratherium acutum

As mentioned in the introduction, a second henkelotheriid, *Drescheratherium acutum* (Fig. 17.8), is present in the coals of the Guimarota mine. This taxon is only known from isolated upper and lower jaws with their dentitions. *Drescheratherium* is mainly characterized by its extremely large, pointed, needle-like canine in the upper jaw. The second upper premolar, which is small in *Henkelotherium*, is completely reduced in this taxon. In its position, a small diastema and a slight constriction in the margin of the jaw is found. In the lower jaws referred to this taxon, the first premolar is reduced, rather than the second as it is the case in *Henkelotherium*. This tooth is very small, or it is missing completely. Thus, the reduction is quite advanced in these animals. The reduction of the first premolar in the lower jaw in *Drescheratherium* might be connected with the development of the very large canine in the upper jaw, which needed some space in the lower jaw when the jaws were in occlusion. The molars of *Drescheratherium* are also different from those of *Henkelotherium*, mainly in terms of their higher and more pointed cusps. Only five molars are present in both the upper and lower jaws.

The function of the large, pointed canine in *Drescheratherium* is not obvious. Such a small animal cannot have been an aggressive predator. The tooth might have played a role in innerspecific social behavior. However, it seems more probable that it served to catch small and agile prey. This indicates a very swift animal, which is in general accordance with the assumption that *Drescheratherium* was arboreal, as it was the case in *Henkelotherium* (KREBS 1998).

References

Henkel, S. & Krebs, B. (1977): Der erste Fund eines Säugetier-Skelettes aus der Jura-Zeit. – Umschau in Wissenschaft und Technik **77**: 217-218.

Krebs, B. (1971): Evolution of the mandible and lower dentition in dryolestids (Pantotheria, Mammalia). – Zoological Journal of the Linnean Society **50**, supplement 1, Early Mammals: 89-102.

– (1991): Das Skelett von *Henkelotherium guimarotae* gen. et sp. nov. (Eupantotheria, Mammalia) aus dem Oberen Jura von Portugal. – Berliner geowissenschaftliche Abhandlungen A **133**: 1-121.

– (1998): *Drescheratherium acutum* gen. et sp. nov., ein neuer Eupantotherier (Mammalia) aus dem Oberen Jura von Portugal. – Berliner geowissenschaftliche Abhandlungen E **28**: 91-111.

Krebs, B. & Hu, Y. (in press): The petrosal of *Henkelotherium* and the evolution of mammalian inner ear. – Journal of Mammalian Evolution.

Taphonomy of the fossil lagerstatte Guimarota

Uwe Gloy

"The transition of an organism from the biosphere into the lithosphere" – this is the essence of the term taphonomy, as coined by EFREMOV (1940). This short definition already shows the broad scope and potential of this rather young branch of paleontology. From the time of the death of an organism, its corpse is subject to a multitude of mechanical, chemical or biological factors. The processes that work on a corpse until it is finally buried in the sediment follow certain patterns. The understanding of these processes is the subject of biostratinomy (WEIGELT 1919), one of the branches of taphonomy.

The processes of deposition and burial in the sediment lead to the next step of the corpse in becoming a fossil. Again, the dead body is subject to mechanical, chemical and biological processes, which, under favorable conditions lead to its fossilization. These processes are called fossil diagenesis.

An important source of information of taphonomical processes are observations in Recent animal corpses in the wild or disarticulation studies in the laboratory. These can than be applied to a fossil lagerstatte, in accordance with the actualistic principle. However, an interpretation of the processes of fossilization is only possible if the geological and paleoecological settings of a fossil locality are more or less known.

Problems of the Guimarota mine

Since the excavations in the Guimarota mine took place in a former commercial coal mine, approximately 80 m underground, the conditions for gathering scientific data were far from ideal. The coal seam, which was usually only 40-60 cm thick, had to be mined in a lying position in the low coal face (Fig. 18.1). Thus, a successive removal of separate layers of coal and a documentation of the detailed orientation of fossils in the sediment, as it is usually carried out in scientific excavations, was not possible. Instead, the entire thickness of the coal seam had to be removed and the fossils were extracted on the surface, where the coal was split and examined by trained workers. Thus, the provenance of the fossils cannot be located within distinct layers, but only roughly within the area that was mined in a day or a week. The lack of this essential information makes a detailed taphonomic investigation difficult. Furthermore, no sedimentological or geological studies of the coal seam in situ are possible today, since the tunnels of the Guimarota mine are no longer accessible.

Despite these unfavorable conditions, a taphonomic investigation of the coals of the Guimarota mine is still possible. The basis for such a taphonomic study are coal-petrographic and geochemical analyses of the borecores, which were extracted during the excavations (GLOY in prep.). Only the interpretation of the resulting information will allow a reconstruction of the formation of the locality. The second important basis for the taphonomic analysis of the lagerstatte is a three-dimensional map of the coal mine, which shows where the fossils, or better, the coal that yielded the fossils, had come from. During the excavations, precise sketches of the excavated areas were drawn each week for eight years. A compos-

Fig. 18.1. Mining the coal seam in the low coal face.

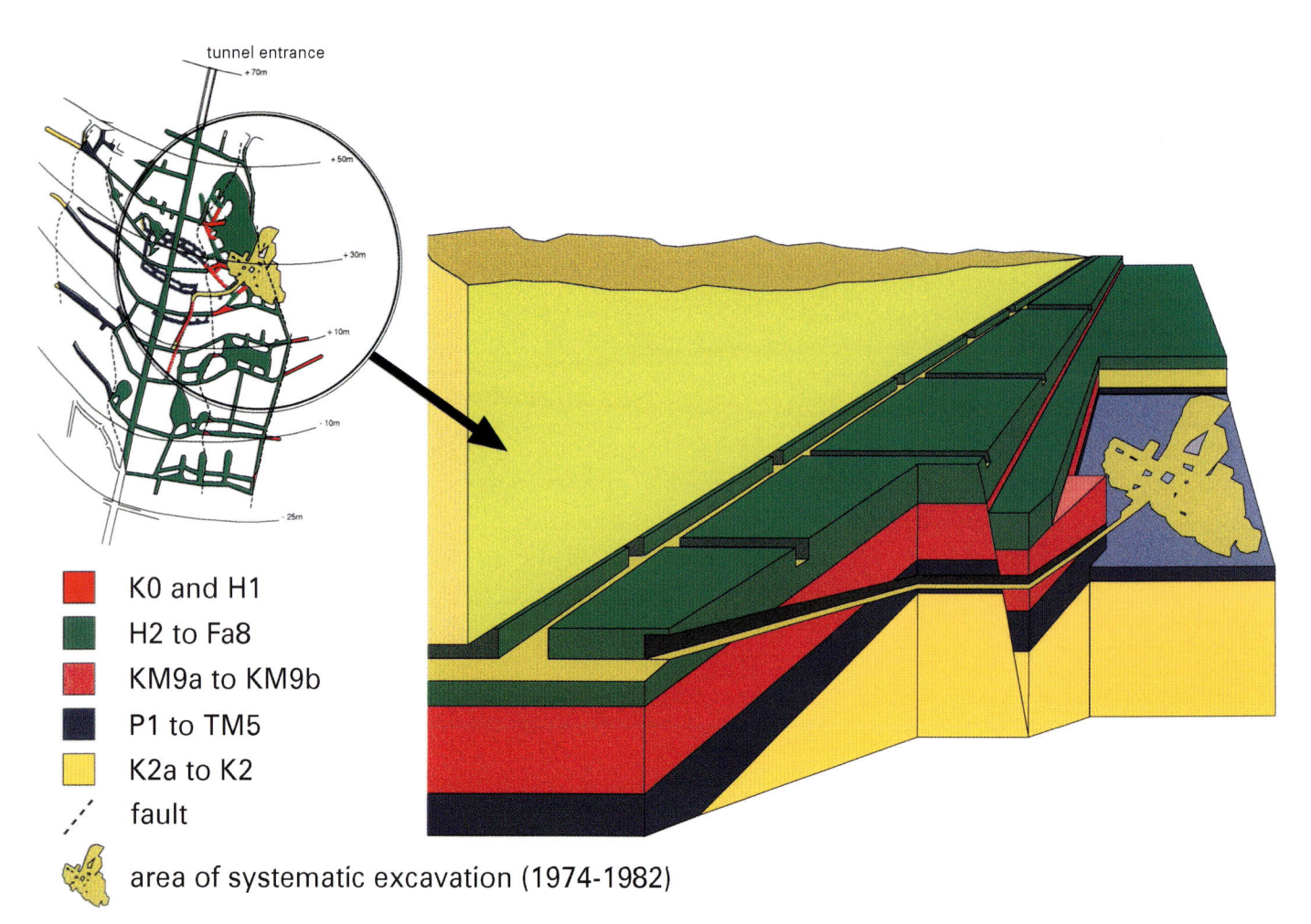

Legend:
- K0 and H1
- H2 to Fa8
- KM9a to KM9b
- P1 to TM5
- K2a to K2
- fault
- area of systematic excavation (1974-1982)

Fig. 18.2. Block diagram of the tunnels of the Guimarota mine (reconstructed from maps of the mine and field notes). For explanations of the lithology of different layers see chapter "Geological setting and dating of the Guimarota beds".

ite map of the excavated area could be reconstructed on the basis of these sketches (Figs. 18.2 and 18.3). The fossils were well labeled during the excavations: each specimen was tentatively identified and labeled with the date it was found. Therefore, each dated fossil from the systematic excavations (10.9.1974–11.6.1982) can be relocated on the map of the excavated area. Thus, it is possible to gain an overview over the relative abundance of the separate group of organisms, although the data must be seen with caution, since the vertical position of the remains within the coal seam remains unknown. Furthermore, with the help of the map it is possible to decide whether bones that were found in isolation might belong to a single individual. No maps exist for fossils that were found before September 10, 1974. The data available for analyses is restricted to the lower coal seam, which was exclusively mined in the systematic excavations. Thus, the taphonomic study is mainly concerned with this lower coal seam.

During the excavations, c. 100 kg of coal were daily dissolved in lye (KREBS 2000*), screen washed using the HENKEL-process (HENKEL 1966), and then the denser phosphatic remains were separated from the residue. The fossils were then picked out and stored in small boxes and laboratory cells. Thus, representative data for the statistical interpretation of the fauna is contained in these so called "GUILHERMECITO-cells" (named after the son of the pithead foreman, who picked out the vertebrate fossils from the residue). Articulated specimens are disarticulated and small fossils may be destroyed during the HENKEL-process; however, the fossils found during splitting the coal show that articulated remains were extremely rare in the first place.

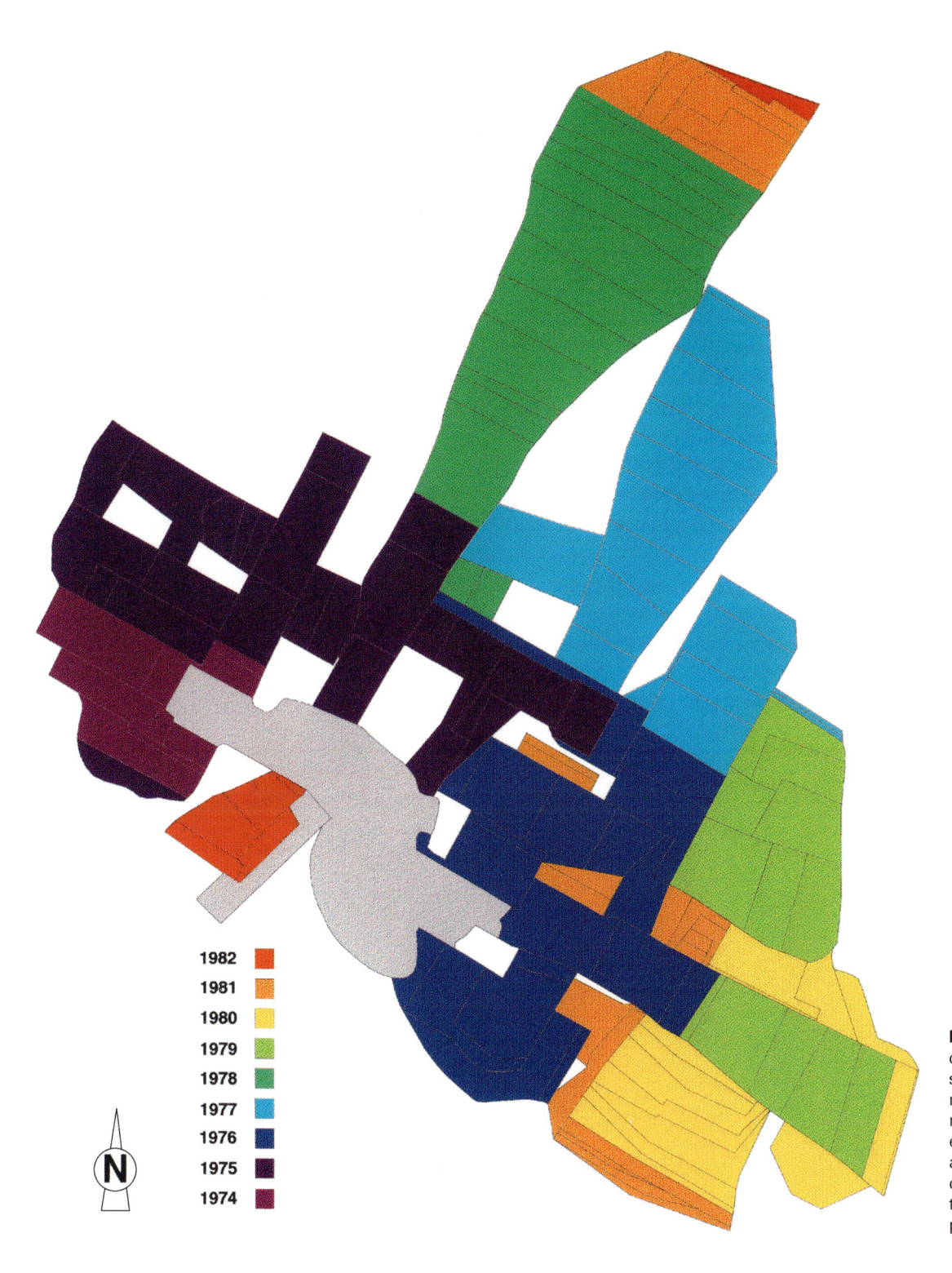

Fig. 18.3. Map of the excavations in the lower coal seam (reconstructed from maps of the mine and field notes). The areas mined each year from 1974-1982 are shown in different colors; the thin lines mark the limit of areas mined per month.

Fig. 18.4. Mangrove at the southern end of the everglades in Florida ("Flamingo Wilderness Waterway").

Fossil diagenesis in brown coal lagerstatten

Some general aspects of the biostratinomy and fossil diagenesis resulting from the geology and the fossils of the Guimarota mine may be discussed before the detailed taphonomic investigation, which is presently being carried out. The potential for preservation of vertebrates and calcareous invertebrates is usually low in coal localities, since humin acids produced by decaying plants usually destroy all phosphatic and carbonatic remains. Thus, the destruction of organic hard parts can only be avoided if the humin acids are neutralized. If the coal seams are flooded by carbonatic water, for example, the lime neutralizes the acids and thus prevents the dissolution of organic phosphatic or calcareous remains. This phenomenon led to the formation of several famous fossil lagerstatten in brown coals. The Eocene brown coals of the Geiseltal, near Halle, for example, are surrounded by extensive Muschelkalk-limestones (Triassic). Here, the carbonatic waters from karst streams and surface drainage neutralized the humin acid (HERRMANN 1930, KRUMBIEGEL 1959). It is thanks to this process that the skeletons of tapirs, early horses, crocodiles, and many other vertebrates are so well preserved in the brown coals of the Geiseltal.

The Recent mangroves and fresh-water swamps of the everglades (Fig. 18.4) are located on marine limestones of a carbonate platform that exists since the Jurassic (RANDAZZO 1997, SCOTT 1997). Here, the swamp deposits are directly neutralized by water dissolving the underlying carbonatic beds. In brackish to marine areas of the mangroves, the submerged plant remains are embedded in a fine mud of lime, which further favors the neutralization of the acids. Own observations showed that mollusk- and vertebrate remains in the Tertiary limestones are very well preserved, although they were covered by plant remains and water that is brown from humin acids for a longer time.

The geological setting of the Guimarota mine is comparable to that of the Eocene coal localities of the Geiseltal and the Recent mangroves of the everglades. In the Guimarota mine, the coal seam is also intercalated in limestones above and below, so that humin acids were neutralized. The neutralization was further enhanced by the fact that the coals are not pure, but rather represent coaly marls petrographically (HELMDACH 1966, 1971).

Preservation of fossils from the Guimarota mine

Most vertebrate remains are not preserved in articulation and they are often fragmentary; articulated remains are extremely rare. The taphonomic processes working on different groups of vertebrates are mainly influenced by the lifestyle and the construction of the skeleton of the animals; thus, animals that lived in the water had a better chance of being preserved in articulation than terrestrial organisms that were washed into the areas of deposition over longer distances. Since the taphonomic analysis of the fossil lagerstatte Guimarota is still work in progress, the results presented here should be seen as tentative.

Fishes

Since sharks (Chondrichthyes) build their cranial and axial skeleton mainly from unossified cartilage, their preservation is usually restricted to mainly isolated teeth, scales, head-, and fin-spines. Thus, several hundreds of shark teeth and scales were found in the residue resulting from the screen washing of the coals of Guimarota. Since sharks replace their teeth throughout their lifetime, it is to be expected that these elements are much more common than head- or fin-spines (KRIWET 2000*). The remaining fish fauna, which is represented by several different groups of Osteichthyes, is also mainly known from isolated teeth and scales, apart from a few isolated skeletal elements. The only exception is the articulated partial skull of a caturid. Since articulated skeletal remains were especially sought during the splitting of the coal, we can assume that the skeletons were disarticulated prior to burial.

Amphibians

The albanerpetontids are one of the most common group of animals in the ecosystem of Guimarota, being represented by thousands of jaw fragments, cranial and postcranial bones. The remains are exclusively disarticulated and mostly fragmented; articulated skeletons have not been found. It is striking that the bones show many breaks, but lack signs of abrasion such as scratches. Thus, prolonged transport within the suspended sediment in flowing waters can be excluded. WIECHMANN (2000*) assumed a burrowing lifestyle, as it is found in Recent salamanders and gymnophionans, for the albanerpetontids, based on anatomical studies and comparisons with modern animals. It therefore seems possible that the amphibians, which lived at the banks of the water, were washed into the area of deposition of Guimarota during floods, together with larger amounts of sediment, in small mud flows. The current disarticulated the decaying corpses of the amphibians and thus led to the embedding of isolated bony elements in the sedimentary mud. Petrographic studies of the coals of Guimarota confirmed that there were times of high sediment influx. If this is really correlated with mass occurrences of albanerpetontid remains is currently being studied (GLOY in prep.); this mode of transport would explain their high abundance in the coals of Guimarota.

Reptiles
(Chelonia, Squamata, Crocodilia, Dinosauria)

The preservation potential of reptiles is mainly determined by their size and lifestyle. Semiaquatic reptiles, such as crocodiles and turtles, have a higher chance of being preserved in articulation than strictly terrestrial forms. The remains of lizards and dinosaurs had to be transported into the area of deposition over a longer way and were thus subject to prolonged mechanical stresses. If we assume a swamp-like habitat, the size of terrestrial animals may be limited. The soft substrate and dense vegetation would represent a severe problem for rapid and safe locomotion of large and heavy animals. According to RAUHUT (2000*), this may be an explanation for the almost complete lack of remains of large dinosaurs. This might, in a limited way, also be true for aquatic animals. Large individuals could only live in open and deep waters. A dense undergrowth formed by roots and plants was a habitat only for small animals.

The mainly disarticulated preservation of the fossils indicates the presence of currents in the area of deposition. First, the bloated corpses drifted on the water surface for a while. In the advanced stages of decay, peripheral parts of the body (skull, tail, limbs) separated from the trunk of the corpse and sank to the bottom of the water. After the remaining parts of the corpse had finally sunk to the bottom and settled there, the current moved parts of the skeleton until they were covered by sediment. Here, the size of the corpses is an important factor, since the skeletal remains of a crocodile of 2 m length (KREBS & SCHWARZ 2000*) are certainly not as easily transported and disarticulated as those of the small lizards (BROSCHINSKI 2000*). Furthermore, small animal corpses are not as rapidly destroyed by scavengers as smaller ones. Therefore, several articulated remains of crocodiles have been found, and the skeleton of *Goniopholis* mentioned above was probably more or less complete when it was buried, but only parts of it could be excavated in the work underground.

In the reptiles, the isolated remains also show very little signs of abrasion, indicating that they were also only transported over short distances, and by weak currents. Since reptiles, like sharks, permanently replace their teeth throughout their lifetime, the high number of reptilian teeth found is not surprising. Especially crocodile teeth have been found in tens of thousands during the screen-washing. Therefore, we can assume that the crocodiles lived in the area of deposition; this is, however, not the case for the marine crocodile *Machimosaurus*, which lived in the open ocean. Since dinosaurs and many lizards were terrestrial animals, only comparatively few corpses got into the coaly swamp. The taphonomy of the small, partially burrowing lizards can be compared with that of the albanerpetontids.

An exception are the turtles, which, in terms of their taphonomy, must be treated separately, because of the anatomical peculiarity of them having a bony shell. Turtles are among the more common faunal elements in the Guimarota mine. A still poorly understood phenomenon is the fact that the turtles are also almost exclusively represented by isolated elements, despite their rather rigid bony shell. According to GASSNER (2000*), this might at least partially reflect the juvenile ontogenetic stage of the turtles, since the separate elements of the shell are not fused in juveniles. Thus, the shell disarticulated relatively rapidly after the death of the animals, the otherwise protected postcranial skeleton was transported together with the separate elements of the theca, and the elements were buried is isolation. Again, the isolated skeletal elements don not exhibit any significant abrasion, indicating a short transport. Unfortunately, the majority of the turtle remains were found in the old excavations, for which only very limited data is available, making their taphonomic interpretation very difficult.

Pterosaurs and Archaeopterygiformes

The development of hollow, thin-walled bones is an adaptation for flight. Such thin bones are rarely preserved as fossils, since they are usually compressed by the weight of the overlying sediment and then destroyed by diagenetic processes after burial. Postcranial remains of pterosaurs from the Guimarota mine are restricted to a few, poorly preserved limb bones and girdle elements (WIECHMANN & GLOY 2000*); their representation by teeth is far better. Due to the permanent tooth replacement of these animals, several hundred of the typical, dagger-like teeth have been preserved.

In contrast, only few teeth of cf. *Archaeopteryx* have been found. Cranial or postcranial skeletal elements of this taxon are missing completely, or have at least not been identified so far.

Mammals

With almost 1000 jaw remains and two skeletons, the Guimarota mine can be regarded as one of the most important localities for early mammals (HAHN & HAHN 2000*, KREBS 2000*, MARTIN 2000*, MARTIN & NOWOTNY 2000*). As it is the case in the other groups of vertebrates, the mammals are mainly represented by isolated elements. Robustly built skeletal elements, such as jaws and few limb bones, are usually well preserved. The question, whether the jaw remains were especially abundant in some areas of the coal is currently being investigated (GLOY in prep.).

The almost complete skeleton of a paurodontid is an exception (KREBS 2000*). In this case, rapid burial must be assumed, which prevented the destruction of the skeleton by currents and decay. Maybe, the corpse was transported into the area of deposition with receding flood waters and there rapidly buried in shallow water. More information on this might be provided by the reconstruction of the original position of the skeleton as it was found, and a petrographic study of the coal (GLOY in prep.).

Paleoecology of the flora and fauna of the Guimarota mine

Looking at the faunal association of the lagerstatte Guimarota in terms of its ecology, the heterogeneity is striking. Apart from terrestrial and limnic organisms, brackish and true marine components are also found in the fauna, which thus comprises animals from different habitats. HELMDACH (1971) assumed the depositional environment to have been a lagoon with fresh water inlets, which was periodically flooded by marine waters. According to BRAUCKMANN (1978), the macroflora indicates a limnic to brackish lagoon. In contrast, the microflora is dominated by terrestrial elements (VAN ERVE & MOHR 1988, MOHR 1989), indicating an upper coastal plain environment. Therefore, the bone bearing coal seam cannot be regarded as a homogeneous unit. Both macro- and microflora only represent a single period within a depositional cycle, since the studied samples were only derived from the upper part (Fa_{11} and Fa_{10}) of the lower coal seam (BRAUCKMANN 1978, VAN ERVE & MOHR 1988). HELMDACH (1968, 1971) analyzed the distinct levels of the coal seam and noticed a change in lithology, which is also reflected by the ostracode faunas described by HELMDACH, but he nevertheless tried to explain the different floras and faunas within a homogeneous habitat.

Preliminary investigations of the coal petrography of bore cores of the lower coal seam show that the different depositional horizons must be analyzed separately (GLOY in prep.). The fossiliferous lower coal seam, which is up to 1m in thickness (usually 40-60 cm; KM_{11}-Fa_{11}, according to HELMDACH 1968), includes a series of sediments that were deposited under different ecological conditions, which were controlled by regressive and transgressive phases of the sea. This assumption is confirmed by the sedimentological studies carried out by HELMDACH (1968), which also show that we are dealing with a nearshore environment. Similar depositional sequences are found in the mangrove swamps of the Ten Thousand Islands (Everglades/Florida), a depositional environment and ecosystem which seems to be well comparable with that of the Guimarota mine. PARKINSON (1989) described the transgressive development of the Ten Thousand Islands within the last 3,000 years. In this geologically rather short period, the mangrove islands exhibit changes from dry terrestrial conditions, mainly limnic influenced swamps, to brackish and even marine conditions. This changing depositional history is documented by only two meters of sediment.

The mining procedures of the coals of the Guimarota mine, in which the whole thickness of the coal seam was removed at the same time, led to a mixture of different ecological horizons. Only detailed petrographical and geochemical investigations of the bore cores and their comparison with similar localities will allow an interpretation of the genesis of the coals and a referral of different faunal elements to distinct ecological horizons (GLOY in prep.). The referral of the different groups of organisms will be based on actualistic comparisons; the swamps and mangroves of the Ten Thousand Islands might provide important information for the interpretation of the ecosystem and paleoecology of the Guimarota mine.

References

EFREMOV, J. A. (1940): Taphonomy: a new branch of geology. – Pan American Geology **74**: 81-93.
BRAUCKMANN, C. (1978): Beitrag zur Flora der Grube Guimarota (Ober-Jura; Mittelportugal). – Geologica et Palaeontologica **12**: 213-222.
BROSCHINSKI, A. (2000*): The lizards from the Guimarota mine. – In: MARTIN, T. & KREBS, B. [eds.] Guimarota – a Jurassic ecosystem: 59-68, München (Verlag Dr. F. Pfeil).
DAVIS, R. A. Jr. (1997): Geology of the Florida Coast. – In: RANDAZZO, A. F. & JONES, D. S. [eds.] The Geology of Florida: 155-168, Gainesville (University Press of Florida).
GASSNER, T. (2000*): The turtles from the Guimarota mine. – In: MARTIN, T. & KREBS, B. [eds.] Guimarota – a Jurassic ecosystem: 55-58, München (Verlag Dr. F. Pfeil).
HAHN, G. & HAHN, R. (2000*): The multituberculates from the Guimarota mine. – In: MARTIN, T. & KREBS, B. [eds.] Guimarota – a Jurassic ecosystem: 97-107, München (Verlag Dr. F. Pfeil).
HELMDACH, F. F. (1966): Stratigraphie und Tektonik der Kohlengrube Guimarota bei Leiria (Mittel-Portugal) und ihrer Umgebung. – Unpublished diploma thesis, Freie Universität Berlin. – 75 pp., Berlin.
– (1968): Oberjurassische Süß- und Brackwasserostracoden der Kohlengrube Guimarota bei Leiria (Mittelportugal). – Unpublished PhD thesis, Freie Universität Berlin. – 92 pp., Berlin.
– (1971): Stratigraphy and ostracode-fauna from the coal mine Guimarota (Upper Jurassic). – Memórias dos Serviços Geológicos de Portugal **17**: 43-88.
HENKEL, S. (1966) Methoden zur Prospektion und Gewinnung kleiner Wirbeltierfossilien. – Neues Jahrbuch für Geologie und Paläontologie, Monatshefte **1966**: 178-184.
HERRMANN, R. (1930): Salzauslaugung und Braunkohlebildung im Geiseltalgebiet bei Merseburg. – Zeitschrift der deutschen geologischen Gesellschaft **82**: 1-85.
KREBS, B. (2000*): The henkelotheriids from the Guimarota mine. – In: MARTIN, T. & KREBS, B. [eds.] Guimarota – a Jurassic ecosystem: 121-128, München (Verlag Dr. F. Pfeil).
KREBS, B. & SCHWARZ, D. (2000*): The crocodiles from the Guimarota mine. – In: MARTIN, T. & KREBS, B. [eds.] Guimarota – a Jurassic ecosystem: 69-74, München (Verlag Dr. F. Pfeil).
KRIWET, J. (2000*): The fish fauna from the Guimarota mine. – In: MARTIN, T. & KREBS, B. [eds.] Guimarota – a Jurassic ecosystem: 41-50, München (Verlag Dr. F. Pfeil).
KRUMBIEGEL, G. (1959): Die tertiäre Pflanzen- und Tierwelt der Braunkohle des Geiseltal. – 156 pp., Wittenberg Lutherstadt (A. Ziemsen Verlag).
MARTIN, T. (2000*): The dryolestids and the small peramurid from the Guimarota mine. – In: MARTIN, T. & KREBS, B. [eds.] Guimarota – a Jurassic ecosystem: 109-120, München (Verlag Dr. F. Pfeil).
MARTIN, T. & NOWOTNY, M. (2000*): The docodont *Haldanodon* from the Guimarota mine. – In: MARTIN, T. & KREBS, B. [eds.] Guimarota – a Jurassic ecosystem: 91-96, München (Verlag Dr. F. Pfeil).
MOHR, B. A. R. (1989): New palynological information on the age and environment of Late Jurassic and Early Cretaceous vertebrate localities of the Iberian Peninsula (eastern Spain and Portugal). – Berliner geowissenschaftliche Abhandlungen A **106**: 291-301.

PARKINSON, R. W. (1989): Decelerating Holocene sea-level rise and its influence on southwest Florida coastal evolution: A transgressive/regressive stratigraphy. – Journal of Sedimentary Petrology **59**: 960-972.

RANDAZZO, A. F. (1997): The Sedimentary Platform of Florida: Mesozoic to Cenozoic. – In: RANDAZZO, A. F. & JONES, D. S. [eds.] The Geology of Florida: 39-56, Gainesville (University Press of Florida).

RAUHUT, O. W. R. (2000*): The dinosaur fauna of the Guimarota mine. – In: MARTIN, T. & KREBS, B. [eds.] Guimarota – a Jurassic ecosystem: 75-82, München (Verlag Dr. F. Pfeil).

SCOTT, T. M. (1997): Miocene to Holocene History of Florida. – In: RANDAZZO, A. F. & JONES, D. S. [eds.] The Geology of Florida: 57-67, Gainesville (University Press of Florida).

VAN ERVE, A. W. & MOHR, B. (1988): Palynological investigation of the Late Jurassic microflora from the vertebrate locality Guimarota coal mine (Leiria, Central Portugal). – Neues Jahrbuch der Geologie und Paläontologie Monatshefte **1988**: 246-262.

VOIGT, E. (1935): Die Erhaltung von Epithelzellen mit Zellkernen, von Chromatophoren und Corium in fossiler Froschhaut aus der mitteleozänen Braunkohle des Geiseltales. – Nova Acta Leopoldina, N.F. **3**: 339-350.

WEIGELT, J. (1919): Geologie und Nordseefauna. – Der Steinbruch **14**: 228-231.

WIECHMANN, M. F. (2000*): The albanerpetontids from the Guimarota mine. – In: MARTIN, T. & KREBS, B. [eds.] Guimarota – a Jurassic ecosystem: 51-54, München (Verlag Dr. F. Pfeil).

WIECHMANN, M. F. & GLOY, U. (2000*): Pterosaurs and urvogels from the Guimarota mine. – In: MARTIN, T. & KREBS, B. [eds.] Guimarota – a Jurassic ecosystem: 83-86, München (Verlag Dr. F. Pfeil).

Preparation of vertebrate fossils from the Guimarota mine

Ellen Drescher

Fig. 19.1. Preparation of vertebrate fossils from the Guimarota mine with a steel needle under a binocular microscope. In the foreground, several plastic bags with unprepared specimens are visible.

Recovering the skeletal remains

The mined coal was carefully split and examined by local female helpers; the bones and teeth that were recognized were marked with chalk circles. Then, mammals and other vertebrates were tentatively identified within the material and separated under the supervision of a scientist. This work was later carried out by a trained local forewoman, so that no scientist had to be at the excavations the whole year round. To keep the coal moist, each specimen that contained a fossil identified as a mammal was wrapped in thin plastic foil. A thick layer of soft paper, with the date of the find written on it, protected the specimen against mechanical damage. Finally, all finds of one day were shrink-wrapped in tough plastic foil. The finds of several days were gathered in a larger plastic bag, together with a label noting the contents of the bag. This material was most important scientifically, and was worked on immediately after it arrived at the preparation laboratory.

Next priority was given to bags labeled "dentitions", which contained jaws of fishes, amphibians and reptiles; these elements are most useful for identifying the animals. Thus, the scientist got a first overview over the diversity of the vertebrates recovered. Another category of finds was labeled "ossos varios" (miscellaneous bones). This included isolated elements of the postcranial skeleton of mammals (if not identified in the field), reptiles, birds, pterosaurs, amphibians, and fishes.

Fig. 19.2. The slab of coal containing the skeleton of *Henkelotherium* before the preparation. In the upper part of the figure, several ribs (arrow) and, to the right of the latter, the right lower jaw are visible. Major parts of the skeleton are covered by a layer of shiny coal (black).

Fig. 19.3. After the removal of the shiny coal, the almost complete skeleton is visible. The side exposed here was then covered with resin, after carefully cleaning and hardening the bones. Then, the specimen was turned over and completely freed of coal from the other side.

Fig. 19.4. The completely prepared skeleton embedded in the artificial resin matrix. The side that was prepared first is covered by the resin; now, the other side is visible, which was hidden in the coal before.

Preparation of isolated bones

As mentioned above, the preparation of mammalian remains was given priority for scientific reasons, and begun as soon as the finds arrived in Berlin. This work is now finished. However, thousands of other finds still await their preparation and scientific investigation.

The preparation of isolated jaws and other bones is relatively simple, it is mainly done with the help of sharply pointed steel needles (DRESCHER 1989) (Fig. 19.1). Highly elastic sewing needles or fine needles for mercerized yarn are suited for this job; they are inserted into the holder with the loop end and then sharpened under the microscope on an oiled grindstone, so that the tip becomes square-headed. Even small needles of only 0.2 mm diameter can be thus sharpened. These are used to clean alveoli or the spaces between the teeth.

The state of preservation of a specimen can be evaluated on the basis of the part that is visible on the slab. If it seems to be robust, the surrounding matrix is simply removed with the needle. If the coal is very marly and thus hard, it proved useful to moisten it with a wet brush. Afterwards, the exposed parts of the fossil are hardened, using Cyanolit 201 of the company BOSTIK. This is a one-component glue that hardens very rapidly ("superglue"), which also invades minute cracks, due to its water-like viscosity. Under the atmospheric humidity, the glue hardens within seconds. Broken and fragile bones were also hardened with this method.

Once the fossil is prepared from one side, the slab containing it is cut around the specimen with a small diamond-rock saw; then the bone or jaw can be completely freed from the matrix. Finally, the fossil is carefully washed with acetone, to remove remaining glue and sediment and thus expose detailed surface structures.

During the splitting of the coal in field, bone fragments had often been broken off, or the bone was split as well. Both halves containing the bone fragments were packed separately. In this case, the fragments had first to be put together again, to get an impression of their original orientation. Since the aim was to free the bones completely from the matrix, the fragments were then prepared separately and, under lucky circumstances, glued back together again. However, parts of bone were often destroyed or missing entirely, so that a reconstruction is only possible with the help of the impression of the element in the sediment. As filling material, either cement or palavit (a material used for casting in dentistry) was used (KÜHNE 1962).

The preparation of articulated skeletons and skeletal remains

One of the most important finds from the Guimarota mine is the almost completely preserved skeleton of *Henkelotherium guimarotae*. The fossil was originally preserved on an approximately fist-sized slab of coal. The unfavorable position of the fossil, between soft brown coal and an overlying piece of brittle shiny coal (vitrinite) (Fig. 19.2), made its preparation more difficult (DRESCHER 1989). After the specimen was unwrapped in the lab, several ribs, the right lower jaw with its dentition, and a compact limb bone were already visible. The preparation procedure for isolated bones described above could not be used for this obviously articulated skeleton. Therefore, a transfer method, as described by KÜHNE (1961, 1962), was used; this method has also proved to be very successful in the preparation of the Early Tertiary vertebrates from the oily shales of Messel, near Darmstadt. First, the side that was covered by the shiny coal was exposed, which was rather difficult. The following problems occurred: first, the heat of the light of the microscope rapidly dried the moist coal. Thus, sizzling and cracking of the coal were the acoustical and optical side effects of the work. Second, the vitrinite firmly stuck to the bones. It could only be removed piece by piece with a razor-sharp needle. Only after the removal of the shiny coal, it became obvious that an almost complete skeleton of an eupantothere was present (Fig. 19.3).

The exposed bones were immediately hardened, using cyanolith. After the preparation of this side (= the later back side) of the specimen was finished, the transfer to an artificial resin matrix was prepared. First, a sheet of aluminum, which had to be approximately 2 cm wider than the slab was high, was put around the specimen and sealed with the help of sellotape. The resulting mould was leveled and fixed in a container filled with sand. To protect the surface from further drying out, and to fill in the cracks resulting from drying out, first a thin layer of resin was poured over the specimen. Thus, gas bubbles, which are otherwise almost unavoidable, could easily escape from this thin layer. On the following day, the whole mould was filled with resin. After the resin had hardened, the soft brown coal of the backside could be cut levelly parallel to the artificial matrix with a small rock saw. The surface of the resin block was then sanded down with wet abrasive paper, and finally polished. Afterwards, the articulated skeleton could be examined through its "glass coffin" through the microscope, which proved to be very helpful for the further preparation.

Before the other side (= the side now exposed) of the specimen could be prepared, it was x-rayed to determine the position of bones that were still embedded in the matrix more precisely. The relatively soft brown coal could be removed in the usual way, and exposed bones were again hardened with cyanolith. After approximately three months, the last piece of coal had been removed, and thus the skeleton of *Henkelotherium* – now on its artificial resin matrix – was completely exposed (Fig. 19.4).

Since thus the coal is replaced by transparent resin, further preparation can always be carried out later, if necessary. Such preparation was for example carried out 20 years after the original preparation of the specimen on the small counterslab of *Henkelotherium*, which was separated from the main part during the splitting of the coal, and had been prepared separately. During a reinvestigation of the specimen, the petrosal had been recognized under the transparent resin layer. The resin could be removed with the help of a small diamond-drill, and thus the entrance of the petrosal with its enclosed cochlea was exposed.

The same method was used to prepare an unfortunately incomplete skeleton of the docodont *Haldanodon*. It has furthermore proved to be very successful in the preparation of skeletal remains and associated specimens of amphibians and reptiles.

References

DRESCHER, E. K. (1989): Bergung und Präparation jurassischer Säugetiere aus der Grube Guimarota, Portugal. – Berliner geowissenschaftliche Abhandlungen A **106**: 29-35.

KÜHNE, W. G. (1961): Präparation von flachen Wirbeltierfossilien auf künstlicher Matrix. – Paläontologische Zeitschrift **35**: 251-252.

– (1962): Präparation von Wirbeltierfossilien aus kolloidalem Gestein. – Paläontologische Zeitschrift **36**: 285-286.

Overview over the Guimarota ecosystem

THOMAS MARTIN

Geology and history of the excavations

The Guimarota mine is the most important fossil lagerstatte of the world for Late Jurassic mammals and other small terrestrial animals. During the ten years of permanent excavations at this site, tens of thousands of bones, jaws, and teeth were recovered, the preparation of which is not finished to the present day. The fossils of the Guimarota mine have substantially improved our knowledge of Late Jurassic faunas.

The coals of the Guimarota mine were deposited in a coastal swamp within the Lusitanian basin. Extensive coastal plains existed on the eastern rim of the opening North Atlantic ocean, on which abundant vegetation flourished. The dense vegetation led to the formation of small peat layers, which later turned into lignites and brown coal seams. These coal seams were mined in several small mines during the first half of the 20th century; however, these mines were only of local importance, and mainly served to provide coal for houses and small factories. One of the last of these coal mines was the Guimarota mine, which was discovered as a fossil locality by WALTER GEORG KÜHNE during prospecting work in the year 1959. During the commercial mining operations in the Guimarota mine, only the slagheaps could be searched for fossils. After the mine was closed down in 1961, KÜHNE carried out some more collection campaigns, until the material of the slagheaps was too weathered to yield further fossils.

This "old excavation" resulted in a respectable collection of mammals and other Late Jurassic vertebrates, which, even at that time, already received international attention and indicated the great potential of this remarkable fossil locality. At the end of the 1960ies and the beginning of the 1970ies, the former assistant professors, BERNARD KREBS and SIEGFRIED HENKEL, developed a plan to reopen the Guimarota mine and mine the coal only for paleontological purposes. The reopening of the mine, which had been closed for more than 10 years, started in the summer of 1972, and the first fossils of the "new excavation" were recovered in 1973. For ten years, until June 1982, the coal of the Guimarota mine was extracted only for paleontological purposes; it was one of the largest projects in the history of paleontology (see chapter KREBS "The excavations in the Guimarota mine").

The coals of Guimarota are marly coals with a high percentage of non inflammable components. Since the most important marine index fossils, ammonites and foraminifera, are missing in the Guimarota beds, the precise stratigraphic dating of the layers is problematic. Within the regional framework, they can be assigned to the Alcobaça Formation, which can be correlated with the Abadia Formation; the latter can be dated with the help of ammonites. For a more precise stratigraphical dating of the beds, charophytes and especially ostracodes have proved to be useful. They indicate a Kimmeridgian age (c. 151-154 million years ago) for the Guimarota beds. The more precise dating as early Kimmeridgian that is found in older publications, cannot be confirmed on the basis of recent research (see chapter SCHUDACK "Geological setting and dating of the Guimarota beds").

Flora and fauna of the Guimarota ecosystem

The plant remains from the Guimarota mine – rare macrofloral remains and an abundant microflora – indicate a subtropical forest-swamp vegetation, with open bodies of water that were lined by dense groves of horsetails *(Equisetites)*. The forests were mainly made up of conifers, with seed ferns and palm ferns forming the undergrowth. Different kinds of ferns and ginkgos grew in higher regions (see chapter MOHR & SCHULTKA "The flora of the Guimarota mine").

The mollusk fauna from the Guimarota mine is poor in terms of taxa, but rich in terms of number of individuals. Bivalves often form extensive shell layers, which mainly consist of the shells of *Isognomon rugosus* and *"Unio"* cf. *alcobacensis*. *Isognomon rugosus* is a dominant faunal element of brackish lagoons and bays within the Lusitanian basin, while *"Unio"* cf. *alcobacensis* might even represent a true fresh-water clam. In contrast, the few species of snails rather indicate a nearshore marine environment, with the ellobioid *Melampoides jurassicus* being interpreted as an amphibious snail living at densely vegetated shores (see chapter ABERHAN, RIEDEL & GLOY "The mollusk fauna of the Guimarota mine").

Ostracodes and charophytes are the most abundant calcareous microfossils in the coals of the Guimarota mine. The ostracode fauna is dominated by limnic and brackish forms, and the Limnocytheridae, which are regarded as typical indicators of brackish conditions, are especially abundant. The ostracodes exhibit close paleobiogeographical relationships with taxa from the Late Jurassic of western North America. Today, the charophytes are also restricted to fresh-water and brackish environments, so that their ecological implications are in general accordance with those of the ostracodes (see chapter SCHUDACK "Ostracodes and charophytes of the Guimarota beds").

The diverse fish fauna of the Guimarota mine is typical for the Late Jurassic in terms of its taxonomic contents. It consists of cartilaginous fishes (Elasmobranchii) and the so-called higher bony fishes (neopterygians). Cartilaginous fishes are mainly represented by hybodont sharks, which are interpreted as euryhaline sharks and thus could tolerate considerable variations in the salinity levels. The bony fishes found in Guimarota also represent taxa that could tolerate variations in salinity levels; the genus *Lepidotes*, for example, is known from many nearshore marine to limnic-brackish sediments in Europe. Strictly non-predatory fishes (e.g. carps) are absent in the Guimarota mine, since these lineages only arose in the Tertiary (see chapter KRIWET "The fish fauna from the Guimarota mine").

The amphibian fauna of the Guimarota mine is dominated by albanerpetontids, small, salamander-like amphibians, which probably had a burrowing lifestyle. They fit well in an ecosystem with dense vegetation and a moist, soft substrate, in which they hunted for arthropods and small mollusks. The scientific investigation of the amphibian remains of Guimarota has only just begun, further taxa can probably be expected among the thousands of amphibian remains (see chapter WIECHMANN "The albanerpetontids from the Guimarota mine").

The lizard fauna of the Guimarota mine is diverse, and includes three genera of skinks (Scincomorpha) and two genera of Anguimorpha, while no geckos could be identified so far. The lizards exhibit a variety of different ecological adaptations: apart from a small, worm-like skink with probably burrowing habits, two gracile gerrhosaurine-like lizards, a medium sized predatory lizard and a large, long-snouted relative of modern monitor lizards were present. Such a high diversity of forms with different adaptations is only possible in a habitat that is favorable for small reptiles, like it was provided by the dense, subtropical vegetation in the surroundings of the coaly swamp of Guimarota (see chapter BROSCHINSKI "The lizards from the Guimarota mine").

Turtles are known from the Guimarota mine only from fragmentary remains, such as isolated shell elements and limb bones. This is rather surprising, since the shell of a turtle is a robust structure, which usually has a high potential of becoming fossilized more or less completely. All turtle remains from the Guimarota mine represent amphibious forms, which lived in rivers and lakes, but also occurred in nearshore brackish habitats. Tortoises are so far only represented by eggshell fragments (see chapter GASSNER "The turtles from the Guimarota mine" and chapter KOHRING "Eggshells from the Guimarota mine").

Crocodiles are known from the Guimarota mine from several tens of thousands of isolated teeth and bones, almost all of which represent rather small, amphibious forms. Only the most abundant taxon, *Goniopholis*, is known from a skeleton. The high abundance of *Goniopholis*-remains indicates that this crocodile lived in the coaly swamps of Guimarota. The other amphibious crocodiles (cf. *Theriosuchus* and cf. *Bernissartia*) possibly lived further inland in small rivers and lakes without marine influence. The genus *Lisboasaurus*, which is known from a few jaws, represents a small, strictly terrestrial crocodile. The large marine crocodile *Machimosaurus*, which is only known from a single specimen, represents a chance visitor in the ecosystem of Guimarota (see chapter KREBS & SCHWARZ "The crocodiles from the Guimarota mine").

Two peculiarities are noteworthy in the dinosaur fauna of the Guimarota mine. The remains (mainly teeth) almost exclusively represent animals of less than one meter body length, but it cannot be said if these animals are juveniles or small taxa, because of the fragmentary preservation. More than 90 % of the teeth represent predatory dinosaurs (theropods), whereas the large herbivores (sauropods) are only represented by five teeth. Large, and thus heavy dinosaurs probably avoided the soft substrate of the ecosystem of Guimarota. The currents of the inlets were probably to weak to transport the corpses or even isolated bones of large dinosaurs into the swamp (see chapter RAUHUT "The dinosaur fauna of the Guimarota mine").

Flying vertebrates are represented in the Guimarota mine mainly by isolated teeth of urvogels (cf. *Archaeopteryx*) and pterosaurs. For the first time, *Archaeopteryx* is found in a locality other than the lithographic limestones of Solnhofen. Apart from teeth, pterosaurs are also represented by a few postcranial elements. However, only one isolated flight phalanx can be identified as representing a rhamphorhynchoid pterosaur on the basis of the presence of a characteristic longitudinal groove, which served as attachment area for the wing membrane (see chapter WIECHMANN & GLOY "Pterosaurs and urvogels from the Guimarota mine").

The Guimarota mine became world-famous mainly for its mammalian remains. The multituberculates represent an extinct side branch of the stem-lineage of mammals, which was very successful in the Mesozoic. The multituberculates from Guimarota are the oldest representatives of this group, which became extinct in the younger Early Tertiary (Oligocene) and thus represents the longest lived order of all mammals. In contrast to all other Mesozoic mammals, the multituberculates were herbivorous, and probably represented the Mesozoic equivalent of the modern rodents. In analogy to rodents, their taxonomic diversity is the highest of all groups of mammals in the Guimarota mine (see chapter HAHN & HAHN "The multituberculates from the Guimarota mine").

The docodonts also represent an extinct lineage of mammals, without living relatives. Apart from numerous jaws and skull remains, *Haldanodon* is also known from a partial skeleton, which indicates that this taxon was a semiaquatic burrower with a desmane-like appearance. *Haldanodon* probably searched for worms and insect-larvae in the soft substrate of the forest swamp of Guimarota. Its teeth are often worn down to small stumps by the abrasive action of the sediment particles it took in together with its prey (see chapter MARTIN & NOWOTNY "The docodont *Haldanodon* from the Guimarota mine").

The most abundant mammals in the Guimarota mine are the dryolestids, small, shrew- to hedgehog-sized insectivorous animals. In contrast to modern mammals, which have three or four true molars, the dryolestids had eight to nine molars. This special adaptation marks the dryolestids as a side branch of the evolutionary lineage leading to modern mammals. The closest relatives of the dryolestids from Guimarota lived in western North America, and thus indicate close faunal relationships to Europe. The extremely rare primitve "peramurid" only has four molars and is thus closer to the modern mammals than the dryolestids (see chapter MARTIN "The dryolestids and the primitive "peramurid" from the Guimarota mine").

The paurodontids (including henkelotheriids) are close relatives of the dryolestids. The almost complete skeleton of *Henkelotherium* belongs to this family of mammals. The skeletal anatomy of *Henkelotherium* indicates that this taxon possibly lived in the brushwood, where it hunted insects and other invertebrates. The nearest kins of *Henkelotherium* are also found in western North America. A close relative of *Henkelotherium* is *Drescheratherium*, which has an enormous canine in the upper jaw. This tooth was possibly used in innerspecific competition (see chapter KREBS "The henkelotheriids from the Guimarota mine").

A taphonomic analysis of the fossil lagerstatte Guimarota is difficult, since the Guimarota beds are no longer accessible. The coal was mined lying in the low coal face, the removal of separate layers of coal was impossible underground. Therefore, the coal samples and the fossils recovered represent a mixture of the varying history of the coaly swamp of Guimarota. This history is characterized by repeated marine ingressions, which periodically resulted in marine conditions. A good model for comparisons with the ecosystem of Guimarota are the mangrove groves of the everglades in Florida, although these modern environments have a completely different flora. The isolated burial of separate vertebrate bones (with few exceptions) indicates that the corpses drifted on the water surface for longer times, until they finally disarticulated. A disarticulation of the skeletons by strong currents can be excluded, since the bones do not exhibit signs of abrasion (see chapter GLOY "Taphonomie of the fossil lagerstatte Guimarota").

The habitat "Guimarota ecosystem"

The multidisciplinary analysis of the Guimarota mine resulted in a rather detailed image of this Late Jurassic terrestrial ecosystem. However, several questions remain, which could not be answered so far. Only a few of these problems, which are mentioned in the different chapters of the book, will be discussed here.

Thus, the almost complete lack of large terrestrial animals (e.g. sauropod dinosaurs), for example, is still an enigma. What exactly were the limiting factors that only allowed the presence of small vertebrates in the ecosystem of Guimarota?

This is an interesting parallel to the Eocene fossil lagerstatte of Messel, near Darmstadt, which has also yielded only small terrestrial animals. In the case of Messel, it is generally assumed that the swampy surroundings of the former forest lake prevented larger animals from getting close to it. This idea might also be true for the coaly swamp of Guimarota.

A further peculiarity of the Guimarota mine is that it is a "mammal-lagerstatte". The excavations resulted in the recovery of more than 800 identifiable skull and jaw remains of mammals – an incredibly high number for Mesozoic mammals. In contrast, the jaw remains of dinosaurs can be counted on the fingers of one hand, and even the c. 750 isolated dinosaur teeth represent a rather moderate number in comparison to the c. 7,000 isolated mammalian teeth. Even if we assume a certain bias in favor of the mammals, which were the main objective of the excavations, this difference is still significant.

The mammals are almost exclusively represented by jaw and skull remains, with the robust lower jaws being by far the most abundant elements. Postcranial remains, even robust limb bones, such as the humerus and femur, are very rare. Again, it is certainly true that jaws with teeth were predominantly collected. However, looking at coal slabs with small and tiny bones and jaws of lizards, amphibians, and other groups, it is obvious that the Portuguese helpers did not miss many vertebrate remains. Thus, obviously more cranial than postcranial remains became fossilized, or the cranial remains were more likely to survive the processes of fossilization.

Many problems of the ecology are also still open, for example the question of how strong the marine influence in the coastal swamp was. How often did marine ingressions occur, and how long was their duration? Bivalves and snails and several fishes indicate at least periods of brackish influence; however, strictly marine organisms, such as ammonites, for example, are missing completely. Which organisms did really live in the ecosystem "coastal swamp" and which were washed in after their death? Does the fossil association of Guimarota represent a thanatocenosis (death association of organisms that also share a habitat in life), or rather a taphocoenosis (an association of animals that were assembled from different habitats after their death)? These questions are the subject of ongoing research at the Freie Universität Berlin, which will result in a still more detailed understanding of the fossil lagerstatte Guimarota.

The present book summarizes the current state of knowledge of the fossil lagerstatte Guimarota. It is the result of an extensive, interdisciplinary cooperation of researchers in geological and biological sciences, representing different areas of vertebrate and invertebrate paleobiology, micropaleontology, paleobotany, geology, stratigraphy, taphonomy, and related research topics.

Bibliography of the Guimarota mine

ELISABETH KREBS

The following list includes all papers that provide information on the history, geological setting, dating, and floral and faunal elements of the Guimarota mine. In cases where the reference to the locality Guimarota is not evident in the title, it is mentioned in a short note. The papers are cited with complete bibliographical information. Secondary literature, such as reviews of publications on Guimarota, or mentioning of the Guimarota mine in faunal lists or of material from Guimarota in text books, faunal lists or similar works, are not listed.

BANDEL, K. (1991): Gastropods from brackish and fresh water of the Jurassic-Cretaceous Transition (a systematic reevaluation). – Berliner geowissenschaftliche Abhandlungen, (A) **134**, pp. 9-55, 7 Pl., 2 Figs.; Berlin.
[First description of two species of snails from the Guimarota mine.]

BRÄM, H. (1973): Chelonia from the Upper Jurassic of Guimarota mine (Portugal). Contribuição para o Conhecimento da Fauna do Kimeridgiano da Mina de Lignito Guimarota (Leiria, Portugal) III Parte, VII. – Memórias dos Serviços geológicos de Portugal, (nova Série) **22**, pp. 135-141, 8 Figs.; Lisbon.

BRAUCKMANN, C. (1978): Beitrag zur Flora der Grube Guimarota (Ober-Jura; Mittel-Portugal). – Geologica et Palaeontologica, **12**, pp. 213-222, 1 Pl., 1 Fig., 2 Tab.; Marburg.

BRINKMANN, W. (1989): Vorläufige Mitteilung über die Krokodilier-Fauna aus dem Ober-Jura (Kimmeridgium) der Kohlengrube Guimarota, bei Leiria (Portugal) und der Unter-Kreide (Barremium) von Uña (Provinz Cuenca, Spanien). – Documenta naturae, **56**, pp. 1-28, 6 Pl., 4 Figs., 1 Tab.; München.

BUSCALIONI, A. D. & ORTEGA, F. (1995): Crocodylomorphs. – II international Symposium on lithographic Limestones, Cuenca 1995, Field Trip Guide Book, Las Hoyas, a lacustrine Konservat-Lagerstätte, pp. 59-61, 2 Figs.; Madrid (Universidad Complutense).
[*Lisboasaurus* from the Guimarota mine, originally described as a lizard by J. SEIFFERT 1973 is interpreted as a crocodile.]

BUSCALIONI, A. D., ORTEGA, F., PÉREZ-MORENO, B. P. & EVANS, S. E. (1996): The Upper Jurassic maniraptoran theropod *Lisboasaurus estesi* (Guimarota, Portugal) reinterpreted as a crocodylomorph. – Journal of vertebrate Paleontology, **16**/2, pp. 358-362, 3 Figs., 1 Tab.; Los Angeles.

BUTLER, P. M. & KREBS, B. (1973): A pantotherian milk dentition. – Paläontologische Zeitschrift, **47**/3-4, pp. 256-258, 1 Fig.; Stuttgart.

DRESCHER, E. K. (1989): Bergung und Präparation jurassischer Säugetiere aus der Grube Guimarota. – Berliner geowissenschaftliche Abhandlungen, (A) **106**, pp. 29-35, 2 Pl.; Berlin.

ENSOM, P. C. & SIGOGNEAU-RUSSELL, D. (1998): New dryolestoid mammals from the basal Cretaceous Purbeck Limestone Group of southern England. – Palaeontology, **41**/1, pp. 35-55, 15 Figs., 2 Tab.; London.
[The systematic position of *Henkelotherium* from the Guimarota mine is discussed.]

ESTES, R. (1981): Gymnophiona, Caudata. – In: P. WELLNHOFER (ed.), Handbook of Palaeoherpetology, Part 2, XV+115 pp., 31 Figs.; Stuttgart/New York (Gustav Fischer).
[Amphibian remains from Guimarota are mentioned and referred to *Albanerpeton megacephalus*.]

– (1983): Sauria terrestria, Amphisbaenia. – In: P. WELLNHOFER (ed.), Handbook of Palaeoherpetology, Part 10A, XXII+249 pp., 69 Figs.; Stuttgart/New York (Gustav Fischer).
[New interpretations of the material from Guimarota that was described as squamates by J. SEIFFERT 1973.]

EVANS, S. E. (1989): New material of *Cteniogenys* (Reptilia: Diapsida; Jurassic) and a reassessment of the phylogenetic position of the genus. – Neues Jahrbuch für Geologie und Paläontologie, Monatshefte **1989**/10, pp. 577-589, 8 Figs.; Stuttgart.
[Some comments on the *Cteniogenys*-material from the Guimarota mine.]

– (1991): A new lizard-like reptile (Diapsida: Lepidosauromorpha) from the Middle Jurassic of England. – Zoological Journal of the Linnean Society, **103**, pp. 391-412, 16 Figs., 1 Tab.; London.
[The new genus of lepidosauromorphs, *Marmouretta*, also occurs in the Guimarota mine.]

FOSSE, G., KIELAN-JAWOROWSKA, Z. & SKAALE, S. G. (1985): The microstructure of tooth enamel in multituberculate mammals. – Palaeontology, **28**/3, pp. 435-449, Pl. 48-50, 1 Fig. 1 Tab.; London.
[Teeth of multituberculates and a docodont from the Guimarota mine are analyzed.]

GALTON, P. M. (1983): The cranial anatomy of *Dryosaurus*, a hypsilophodontid dinosaur from the Upper Jurassic of North America and East Africa, with a review of hypsilophodontids from the Upper Jurassic of North America.– Geologica et Palaeontologica **17**: pp. 207-243; 4 Pl., 12 Figs., 3 Tab., Marburg.
[A new diagnosis of *Phyllodon* is given.]

GASSNER, T. (1999): Die Schildkröten aus dem Oberen Jura der Kohlengrube Guimarota (Portugal). – Unpublished diploma thesis, Fachbereich Geowissenschaften, Freie Universität Berlin, 48 pp., 4 Pl., 23 Figs.; Berlin

HAHN, G. (1969): Beiträge zur Fauna der Grube Guimarota Nr. 3. Die Multituberculata. – Palaeontographica, (A) **133**/1-3, 100 pp., 10 Pl., 85 Figs., 20 Tab.; Stuttgart.

– (1971): The dentition of the Paulchoffatiidae (Multituberculata, Upper Jurassic). Contribuição para o Conhecimento da Fauna do Kimeridgiano da Mina de Lignito Guimarota (Leiria, Portugal) II Parte, III. – Memórias dos Serviços geológicos de Portugal, (nova Série) **17**, pp. 7-39, 23 Figs., 4 Tab.; Lisbon.

– (1973): Neue Zähne von Haramiyiden aus der deutschen Ober-Trias und ihre Beziehungen zu den Multituberculaten. – Palaeontographica, (A) **142**/1-3, pp. 1-15, 13 Figs.; Stuttgart.

– (1977a): Das Coronoid der Paulchoffatiidae (Multituberculata; Ober-Jura). – Paläontologische Zeitschrift, **51**/3-4, pp. 246-253, Pl.23, 5 Figs.; Stuttgart.

– (1977b): Neue Schädel-Reste von Multituberculaten (Mamm.) aus dem Malm Portugals. – Geologica et Palaeontologica, **11**, pp.161-186, 3 Pl., 11 Figs., 2 Tab.; Marburg.

– (1978a): Die Multituberculata, eine fossile Säugetier-Ordnung. – Naturwissenschaftlicher Verein Hamburg, Sonderband **3**, pp. 61-95, 16 Figs., 1 Tab.; Hamburg.

– (1978b): Milch-Bezahnungen von Paulchoffatiidae (Multituberculata; Ober-Jura). – Neues Jahrbuch für Geologie und Paläontologie, Monatshefte **1978**/1, pp. 25-34, 5 Figs., 1 Tab.; Stuttgart.

– (1978c): Neue Unterkiefer von Multituberculaten aus dem Malm Portugals. – Geologica et Palaeontologica, **12**, pp. 177-212, 5 Pl., 11 Figs., 3 Tab.; Marburg.

– (1981): Zum Bau der Schädel-Basis bei den Paulchoffatiidae (Multituberculata; Ober-Jura). – Senckenbergiana lethaea, **61**/3-6, pp. 227-245, 1 Pl., 4 Figs., 1 Tab.; Frankfurt a.M.

– (1985): Zum Bau des Infraorbital-Foramens bei den Paulchoffatiidae (Multituberculata, Ober-Jura). – Berliner geowissenschaftliche Abhandlungen, (A) **60**, pp. 5-27, 3 Pl., 17 Figs., 1 Tab.; Berlin.

– (1987): Neue Beobachtungen zum Schädel- und Gebiss-Bau der Paulchoffatiidae (Multituberculata, Ober-Jura). – Palaeovertebrata, **17**/4, pp. 155-196, 5 Pl., 8 Figs., 2 Tab.; Montpellier.

– (1988): Die Ohr-Region der Paulchoffatiidae (Multituberculata, Ober-Jura). – Palaeovertebrata, **18**/3, pp. 155-185, 11 Pl., 1 Fig., 2 Tab.; Montpellier.

– (1993): The systematic arrangement of the Paulchoffatiidae (Multituberculata) revisited. – Geologica et Palaeontologica, **27**, pp. 201-214, 2 Figs., 4 Tab.; Marburg.

HAHN, G. & HAHN, R. (1994): Nachweis des Septomaxillare bei *Pseudobolodon krebsi* n. sp. (Multituberculata) aus dem Malm Portugals. – Berliner geowissenschaftliche Abhandlungen, (E) **13**, pp. 9-29, 3 Pl., 5 Figs., 2 Tab.; Berlin.

– (1998a): Neue Beobachtungen an Plagiaulacoidea (Multituberculata) des Ober-Juras. 1. Zum Zahn-Wechsel bei *Kielanodon*. – Berliner geowissenschaftliche Abhandlungen, (E) **28**, pp. 1-7, 1 Pl., 1 Fig.; Berlin.

– (1998b): Neue Beobachtungen an Plagiaulacoidea (Multituberculata) des Ober-Juras. 2. Zum Bau des Unterkiefers und des Gebisses bei *Meketibolodon* und bei *Guimarotodon*. – Berliner geowissenschaftliche Abhandlungen, (E) **28**, pp. 9-37, 3 Pl., 33 Figs., 4 Tab.; Berlin.

– (1998c): Neue Beobachtungen an Plagiaulacoidea (Multituberculata) des Ober-Juras. 3. Der Bau der Molaren bei den Paulchoffatiidae. – Berliner geowissenschaftliche Abhandlungen, (E) **28**, pp. 39-84, 52 Figs., 13 Tab.; Berlin.

– (1998d): Neue Beobachtungen an Plagiaulacoidea (Multituberculata) des Ober-Juras. 4. Ein Vertreter der Albionbaataridae im Lusitanien Portugals. – Berliner geowissenschaftliche Abhandlungen, (E) **28**, pp. 85-89, 2 Figs.; Berlin.

HELMDACH, F. F. (1966): Stratigraphie und Tektonik der Kohlengrube Guimarota bei Leiria (Mittelportugal) und ihrer Umgebung. – Unpublished diploma thesis, mathematisch-naturwissenschaftliche Fakultät, Freie Universität Berlin, 75 pp., 16 Figs., 1 map; Berlin.
- (1968): Oberjurassische Süss- und Brackwasserostrakoden der Kohlengrube Guimarota bei Leiria (Mittelportugal). – Unpublished PhD-thesis, Freie Universität Berlin, 91 pp., 27 Figs., 1 Tab., 1 geological sections, 2 maps; Berlin.
- (1971a): Stratigraphy and ostracode-fauna from the coalmine Guimarota (Upper Jurassic). Contribuição para o Conhecimento da Fauna do Kimeridgiano da Mina de Lignito Guimarota (Leiria, Portugal) II Parte, IV. – Memórias dos Serviços geológicos de Portugal, (nova Série) **17**, pp. 41-88, 4 Pl., 20 Figs., 4 Tab.; Lisbon.
- (1971b): Zur Gliederung limnisch-brackischer Sedimente des portugiesischen Oberjura (ob. Callovien – Kimmeridge) mit Hilfe von Ostrakoden. – Neues Jahrbuch für Geologie und Paläontologie, Monatshefte **1971**/11, pp. 645-662, 14 Figs., 1 Tab.; Stuttgart.
- (1973): A contribution to the stratigraphical subdivision of nonmarine sediments of the Portugese Upper Jurassic. – Comunicações dos Serviços geológicos de Portugal, **57**, pp. 1-21, 4 pl., 6 Figs., 1 Tab.; Lisbon.

HENKEL, S. (1966): Methoden zur Prospektion und Gewinnung kleiner Wirbeltierfossilien. – Neues Jahrbuch für Geologie und Paläontologie, Monatshefte **1966**/3, pp. 178-184, 2 Figs.; Stuttgart.
- (1973): Die Entstehungsgeschichte der Säugetiere. – Pressedienst Wissenschaft FU [Freie Universität] Berlin, **1973**/7, 31 pp., 8 Figs.; Berlin.

HENKEL, S. & KREBS, B. (1977): Der erste Fund eines Säugetier-Skelettes aus der Jura-Zeit. – Umschau in Wissenschaft und Technik, **77**/7, pp. 217-218, 2 Figs.; Frankfurt a.M.

HENKEL, S. & KRUSAT, G. (1980): Die Fossil-Lagerstätte in der Kohlengrube Guimarota (Portugal) und der erste Fund eines Docodontiden-Skelettes. – Berliner geowissenschaftliche Abhandlungen, (A) **20**, pp. 209-216, 4 Figs.; Berlin.

HENKEL, S. & KÜHNE, W. G. (1966): Methoden und Ergebnisse der Guimarotagrabungen (Kimmeridge, Mittelportugal 1959-1963). – Zeitschrift der deutschen geologischen Gesellschaft, **116** (Jahrg.1964)/3, p. 972; Hannover.

HOFFSTETTER, R. (1964): Les Sauria du Jurassique supérieur et spécialement les Gekkota de Bavière et de Mandchourie. – Senckenbergiana biologica, **45**/3-5, pp. 281-324, Taf. 3-10, 4 Figs.; Frankfurt a.M.
[The occurrence of skinkomorphs in the Guimarota mine is mentioned for the first time.]

KOHRING, R. (1990): Upper Jurassic chelonian eggshell fragments from the Guimarota mine (Central Portugal). – Journal of Vertebrate Paleontology, **10**/1, pp. 128-130, 1 Fig.; Los Angeles.
- (1998): Neue Schildkröten-Eischalen aus dem Oberjura der Grube Guimarota (Portugal). – Berliner geowissenschaftliche Abhandlungen, (E) **28**, pp. 113-117, 1 Pl.; Berlin.

KRAUSE, D. W. & HAHN, G. (1990): Systematic position of the Paulchoffatiinae (Multituberculata, Mammalia). – Journal of Paleontology, **64**/6, pp. 1051-1054, 3 Figs.; Tulsa.

KREBS, B. (1967): Der Jura-Krokodilier *Machimosaurus* H. V. MEYER. – Paläontologische Zeitschrift, **41**/1-2, pp. 46-59, 4 Figs.; Stuttgart.
- (1968): Le crocodilien *Machimosaurus*. Contribuição para o Conhecimento da Fauna do Kimeridgiano da Mina de Lignito Guimarota (Leiria, Portugal) I Parte, II. – Memórias dos Serviços geológicos de Portugal, (nova Série) **14**, pp. 21-53, 18 Figs., 5 Tab.; Lisbon.
- (1969): Nachweis eines rudimentären Coronoids im Unterkiefer der Pantotheria (Mammalia). – Paläontologische Zeitschrift, **43**/1-2, pp. 57-63, Pl. 4, 4 Figs.; Stuttgart.
- (1971): Evolution of the mandible and lower dentition in dryolestids (Pantotheria, Mammalia). – Zoological Journal of the Linnean Society, **50**, Supplement No.1: Early mammals, pp. 89-102, 2 Pl., 9 Figs.; London.
- (1975): Zur frühen Geschichte der Säugetiere. – Natur und Museum, **105**/5, pp. 147-155, 3 Figs.; Frankfurt a.M.
- (1980): The search for Mesozoic mammals in Spain and Portugal. – Mesozoic vertebrate Life, **1**/1, pp. 23-25; San Diego.
- (1983) Institut für Paläontologie, Freie Universität Berlin. – News Bulletin, Society of Vertebrate Paleontology, **129**, pp. 37-39; Gainesville.
[Report on the excavations in the Guimarota mine.]
- (1987): The skeleton of a Jurassic eupantothere and the arboreal origin of modern mammals. – Fourth Symposium on Mesozoic terrestrial Ecosystems, Drumheller 1987, Short Papers, pp. 132-137, 3 Figs.; Drumheller (Tyrrell Museum of Paleontology).
- (1988): Mesozoische Säugetiere – Ergebnisse von Ausgrabungen in Portugal. – Sitzungsberichte der Gesellschaft naturforschender Freunde zu Berlin, (neue Folge) **28**, pp. 95-107; Berlin.

- (1991): Das Skelett von *Henkelotherium guimarotae* gen. et sp. nov. (Eupantotheria, Mammalia) aus dem Oberen Jura von Portugal. – Berliner geowissenschaftliche Abhandlungen, (A) **133**, 110 pp., 5 Pl., 12 Figs., 4 Tab.; Berlin.
- (1998): *Drescheratherium acutum* gen. et sp. nov., ein neuer Eupantotherier (Mammalia) aus dem Oberen Jura von Portugal. – Berliner geowissenschaftliche Abhandlungen, (E) **28**, pp. 91-111, 1 Pl., 2 Figs., 1 Tab.; Berlin.

KREBS, B. & HU, Y. (in press): The petrosal of *Henkelotherium* and the evolution of mammalian inner ear. – Journal of mammalian Evolution; New York/London.

KRIWET, J. (1995): Beitrag zur Kenntnis der Fisch-Fauna des Ober-Jura (unteres Kimmeridge) der Kohlengrube Guimarota bei Leiria, Mittel-Portugal: 1. *Asteracanthus biformatus* n. sp. (Chondrichthyes: Hybodontoidea). – Berliner geowissenschaftliche Abhandlungen, (E) **16**, pp. 683-691, 1 Pl., 3 Figs., 1 Tab.; Berlin.
- (1997a): Beitrag zur Kenntnis der Fischfauna des Oberjura (unteres Kimmeridgium) der Kohlengrube Guimarota bei Leiria, Mittel-Portugal: 2. Neoselachii (Pisces, Elasmobranchii). – Berliner geowissenschaftliche Abhandlungen, (E) **25**, pp. 293-301, 2 Pl.; Berlin.
- (1997b): Late Jurassic fishes (Pisces: Elasmobranchii, Actinoperygii) of the Iberian peninsula (preliminary report). – Comunicaciones IV Congreso de Jurásico de España, Alcañiz 1997, pp. 89-90, 1 Fig.; Alcañiz (Ayuntamiento).

KRUSAT, G. (1973): *Haldanodon exspectatus* KÜHNE & KRUSAT 1972 (Mammalia, Docodonta). – Unpublished PhD-thesis, Freie Universität Berlin, 158 pp., 32 Fig., 1 Tab.; Berlin.
- (1980): *Haldanodon exspectatus* KÜHNE & KRUSAT 1972 (Mammalia, Docodonta). Contribuição para o Conhecimento da Fauna do Kimeridgiano da Mina de Lignito Guimarota (Leiria, Portugal) IV Parte. – Memórias dos Serviços geológicos de Portugal, **27**, 79 pp., 12 Pl., 32 Figs., 1 Tab.; Lisbon.
- (1991): Functional morphology of *Haldanodon exspectatus* (Mammalia, Docodonta) from the Upper Jurassic of Portugal. – Fifth Symposium on Mesozoic terrestrial Ecosystems and Biota, Oslo 1991, extended Abstracts, Contributions from the paleontological Museum University of Oslo, No. **364**, pp. 37-38, 1 Fig., Oslo.

KÜHNE, W. G. (1961a): A mammalian fauna from the Kimmeridgian of Portugal. – Nature, **192**/4799, pp. 274-275, 2 Figs.; London.
- (1961b): Eine Mammaliafauna aus dem Kimeridge Portugals. Vorläufiger Bericht. – Neues Jahrbuch für Geologie und Paläontologie, Monatshefte **1961**/7, pp. 374-381, 4 Figs.; Stuttgart.
- (1961c): Une faune de mammiferes lusitaniens (rapport provisoire). – Comunicações dos Serviços geológicos de Portugal, **45**, pp. 211-221, 4 Figs.; Lisbon.
- (1962): Präparation von Wirbeltierfossilien aus kolloidalem Gestein. Ein Behälter für kleine Fossilien. – Paläontologische Zeitschrift, **36**/3-4, pp. 285-286; Stuttgart.
- (1963): Oberjura-Mammalia aus Portugal. – Paläontologische Zeitschrift, **37**/1-2, p. 15; Stuttgart.
- (1967): Ursprung und Entwicklung der Säugetiere. – Umschau in Wissenschaft und Technik, **67**/9, pp. 288-289, 1 Figs.; Frankfurt a.M.
- (1968a): History of discovery, report on the work performed, procedure, technique and generalities. Contribuição para o Conhecimento da Fauna do Kimeridgiano da Mina de Lignito Guimarota (Leiria, Portugal) I Parte, I. – Memórias dos Serviços geológicos de Portugal, (nova Série) **14**, pp. 7-20, 7 Figs.; Lisbon.
- (1968b): Kimeridge mammals and their bearing on the phylogeny of the Mammalia. – In: E. T. DRAKE (Ed.), Evolution and environment, Article 3, pp. 109-123, 8 Figs.; New Haven/London (Yale University Press).
- (1969): Säugetiere im Schatten der Dinosaurier. – Umschau in Wissenschaft und Technik, **69**/12, pp. 373-377, 7 Figs.; Frankfurt a.M.

KÜHNE, W. G. & KRUSAT, G. (1972): Legalisierung des Taxon *Haldanodon* (Mammalia, Docodonta). – Neues Jahrbuch für Geologie und Paläontologie, Monatshefte **1972**/5, pp. 300-302; Stuttgart.

LESTER, K. S. & KOENIGSWALD, W. V. (1989): Crystallite orientation discontinuities and the evolution of mammalian enamel – or, when is a prism? – Scanning Microscopy, **3**/2, pp. 645-663, 40 Figs.; Chicago.
[Enamel structures of *Haldanodon* and a eupantothere from the Guimarota mine are analyzed and figured.]

LILLEGRAVEN, J. A. & HAHN, G. (1993): Evolutionary analysis of the middle and inner ear of Late Jurassic multituberculates. – Journal of mammalian Evolution, **1**/1, pp. 47-74, 7 Figs., 1 Tab.; New York/London.

LILLEGRAVEN, J. A. & KRUSAT, G. (1991): Craniomandibular anatomy of *Haldanodon exspectatus* (Docodonta; Mammalia) from the Late Jurassic of Portugal and its implications to the evolution of mammalian characters. – Contributions to Geology, University of Wyoming, **28**/2, pp. 39-138, 17 Figs., 1 Tab.; Laramie.

MARTIN, T. (1995): Dryolestidae from the Kimmeridge of the Guimarota coal mine (Portugal) and their implications for dryolestid systematics and phylogeny. – Sixth Symposium on Mesozoic terrestrial Ecosystems and Biota, Beijing 1995, Short Papers, pp. 229-231, 1 Fig.; Beijing (China Ocean Press).

– (1997a): Schmelzmikrostrukturen bei Säugetieren. – In: K. W. ALT & J. C.TÜRP (Eds.), Die Evolution der Zähne, 4/3, pp. 401-422, 20 Figs.; Berlin (Quintessenz Verlag).
[Also concerns dryolestids from the Guimarota mine.]

– (1997b): Tooth replacement in Late Jurassic Dryolestidae (Eupantotheria, Mammalia). – Journal of mammalian Evolution, 4/1, pp. 1-18, 8 Figs., 1 Tab.; New York/London.

– (1999): Dryolestidae (Dryolestoida, Mammalia) aus dem Oberen Jura von Portugal. – Abhandlungen der Senckenbergischen naturforschenden Gesellschaft 550, pp.1-119; Frankfurt a.M.

– (in press): New stem-lineage representatives of Zatheria (Mammalia) from the Late Jurassic of Portugal. – Journal of vertebrate Paleontology; Lawrence.

– (in press): The mammal fauna of the Late Jurassic Guimarota ecosystem. – Ameghiniana; Buenos Aires.

MCGOWAN, G. J. (1998): Frontals as diagnostic indicators in fossil albanerpetontid amphibians. – Bulletin of the national Science Museum, Series C 24/3-4, pp. 185-194, 4 Figs., 1 Tab.; Tokyo. [Also concerns material from the Guimarota mine.]

MILNER, A. R. & EVANS, S. E. (1991): The Upper Jurassic diapsid *Lisboasaurus estesi* – a maniraptoran theropod. – Palaeontology, 34/3, pp.503-513, 7 Figs.; London.
[*Lisboasaurus* from the Guimarota mine is interpreted as a theropod dinosaur.]

MOHR, B. A. R. (1989): New palynological information on the age and environment of Late Jurassic and Early Cretaceous vertebrate localities of the Iberian peninsula (eastern Spain and Portugal). – Berliner geowissenschaftliche Abhandlungen, (A) 106, pp.291-301, 1 Pl., 2 Figs.; Berlin.

MOHR, B. A. R. & SCHMIDT, D. (1988): The Oxfordian/Kimmeridgian boundary in the region of Porto de Mós (Central Portugal): stratigraphy, facies and palynology. – Neues Jahrbuch für Geologie und Paläontologie, Abhandlungen 176/2, pp. 245-267, 10 Figs.; Stuttgart.

POOLE, D. F. G. (1971): An introduction to the phylogeny of calcified tissues. – In: A. A. DAHLBERG (Ed.), Dental morphology and evolution, Part 2, Chapter 6, pp. 65-79, 9 Figs.; Chicago/London (University of Chicago Press).
[Figs. 6 and 7 are thin sections of teeth of dryolestids, which obviously come from the Guimarota mine.]

PROTHERO, D. R. & ESTES, R. (1980): Late Jurassic lizards from Como Bluff, Wyoming and their palaeobiogeographic significance. – Nature, 286/5772, pp. 484-486, 1 Fig.; London.
[Close links between the squamate faunas of the Morrison Formation and the Guimarota mine are pointed out.]

SANDER, P. M. (1997): Non-mammalian synapsid enamel and the origin of mammalian enamel prisms: The bottom-up perspective. – In: W. V. KOENIGSWALD & P. M. SANDER (Eds.), Tooth enamel microstructure, chapter 3, pp. 41-62, 4 Figs.; Rotterdam (Balkema).
[Also concerns mammals from the Guimarota mine.]

SCHMIDT, D. (1982): Zur Geologie des Jura im Gebiet südlich von Porto de Mós (Mittelportugal). – Unpublished diploma thesis, Fachbereich Geowissenschaften, Freie Universität Berlin, 191 pp., 64 Figs., 9 Tab., 3 maps; Berlin.

– (1986): Petrographische und biofazielle Untersuchungen an oberjurassischen Deckschichten des Diapirs von Porto de Mós (Mittelportugal). – Berliner geowissenschaftliche Abhandlungen, (A) 77, 172 pp., 20 Pl., 65 Figs.; Berlin.

SCHUDACK, M. E. (1987): Charophytenflora und fazielle Entwicklung der Grenzschichten mariner Jura / Wealden in den Nordwestlichen Iberischen Ketten (mit Vergleichen zu Asturien und Kantabrien). – Palaeontographica, (B) 204, S. 1-180, 9 Pl., 106 Figs., Stuttgart.

– (1993): Charophyten aus dem Kimmeridgium der Kohlengrube Guimarota (Portugal). Mit einer eingehenden Diskussion zur Datierung der Fundstelle. – Berliner geowissenschaftliche Abhandlungen, (E) 9, pp. 211-231, 1 Pl., 6 Figs.; Berlin.

– (1993): Die Charophyten in Oberjura und Unterkreide Westeuropas. Mit einer phylogenetischen Analyse der Gesamtgruppe. – Berliner geowissenschaftliche Abhandlungen, (E) 8, 209 pp., 20 Pl., 70 Figs., Berlin.

SCHUDACK, M. E., TURNER, C. E. & PETERSON, F. (1998): Biostratigraphy, paleoecology and biogeography of charophytes and ostracodes from the Upper Jurassic Morrison Formation, Western Interior, USA. – Modern Geology, 22, pp. 379-414, 3 Taf., 9 Fig., 1 Tab.; London.
[The most abundant species of ostracodes

from the Guimarota mine is referred to *Timiriasevia guimarotensis*.]

SCHUDACK, U. (1989): Zur Systematik der oberjurassischen Ostracodengattung *Cetacella* MARTIN 1958 (Syn. *Leiria* HELMDACH 1971). – Berliner geowissenschaftliche Abhandlungen, (A) **106**, pp. 459-471, 1 Pl., 2 Figs.; Berlin.

SCHWARZ, D. (1999): Das Skelett eines Goniopholididen (Crocodilia) aus dem Oberen Jura von Portugal. – Unpublished diploma thesis, Fachbereich Geowissenschaften, Freie Universität Berlin, 100 pp., 7 Pl., 33 Fig., 2 Tab.; Berlin.

SCHWIETZER, C.-A. (1981): Trennung, Abbildung und Klassifikation von merkmalsarmen Ostrakoden-Populationen mit Hilfe von Fourier-Koeffizienten, exemplarisch dargestellt an der Gattung *Darwinula* BRADY & ROBERTSON 1885. – Berliner geowissenschaftliche Abhandlungen, (A) **32**, pp. 135-210, 3 Pl., 15 Figs., 6 Tab.; Berlin.
[The genus *Darwinula* from the Guimarota mine is taken into consideration.]

SEIFFERT, J. (1970): Oberjurassische Lazertilier aus der Kohlengrube Guimarota bei Leiria (Mittelportugal). – Unpublished PhD thesis, Freie Universität Berlin, 180 pp., 77 Fig., tables, 1 geological section; Berlin.

– (1973): Upper Jurassic lizards from Central Portugal. Contribuição para o Conhecimento da Fauna do Kimeridgiano da Mina de Lignito Guimarota (Leiria, Portugal) III Parte, V. – Memórias dos Serviços geológicos de Portugal, (nova Série) **22**, pp. 7-85, 4 Pl., 61 Figs., 1 Tab.; Lisbon.

SOMMERSBERG, B. (1996): Ein albanerpetontides Amphib aus dem Oberjura der Kohlengrube Guimarota (bei Leiria, Portugal). – Unpublished diploma thesis, Fachbereich Geowissenschaften, Freie Universität Berlin, 69 pp., 16 Figs., 1 Tab., 6 maps; Berlin.

THULBORN, R. A. (1973): Teeth of ornithischian dinosaurs from the Upper Jurassic of Portugal. Contribuição para o Conhecimento da Fauna do Kimeridgiano da Mina de Lignito Guimarota (Leiria, Portugal) III Parte, VI. – Memórias dos Serviços geológicos de Portugal, (nova Série) **22**, pp. 89-134, 27 Figs.; Lisbon.

VAN ERVE, A. & MOHR, B. A. R. (1988): Palynological investigations of the Late Jurassic microflora from the vertebrate locality Guimarota coal mine (Leiria, Central Portugal). – Neues Jahrbuch für Geologie und Paläontologie, Monatshefte **1988**/4, pp. 246-262, 5 Figs., 2 Tab.; Stuttgart.

WEIGERT, A. (1995): Isolierte Zähne von cf. *Archaeopteryx* sp. aus dem Oberen Jura der Kohlengrube Guimarota (Portugal). – Neues Jahrbuch für Geologie und Paläontologie, Monatshefte **1995**/9, pp. 562-576, 11 Figs.; Stuttgart.

ZINKE, J. (1996): Die Theropoden-Zähne aus dem Oberen Jura der Grube Guimarota bei Leiria (Portugal). – Unpublished diploma thesis, Fachbereich Geowissenschaften, Freie Universität Berlin, 58 pp., 11 Pl., 15 Figs., 9 Tab., Appendix 1 with 1 Tab., Appendix 2 with 6 Tab.; Berlin.

– (1998): Small theropod teeth from the Upper Jurassic coal mine of Guimarota (Portugal). – Paläontologische Zeitschrift, **72**/1-2, pp. 179-189, 8 Figs., 1 Tab.; Stuttgart.

ZINKE, J. & RAUHUT, O. W. M. (1994): Small theropods (Dinosauria, Saurischia) from the Upper Jurassic and Lower Cretaceous of the Iberian peninsula. – Berliner geowissenschaftliche Abhandlungen, (E) **13**, pp. 163-177, 2 Pl., 4 Figs.; Berlin.

Floral and faunal list of the Guimarota mine

Plantae

Pteridophyta
  Equisetales
    Equisetaceae
      Equisetites lusitanicum HEER 1881
      Calamospora mesozoica COUPER 1958
  Lycopodiatae
    Lycopodiaceae
      Retitriletes clavatoides (COUPER 1958) DÖRING et al. 1963
  Filicatae
    Osmundaceae
      Osmundacites wellmannii COUPER 1953
    Gleicheniaceae
      Gleicheniidites senonicus ROSS 1949
    Cyatheaceae
      Deltoidospora sp.
    Schizaeaceae
      Ischyosporites marburgensis DE JERSEY 1953
    Matoniaceae
      Deltoidospora mesozoica (COUPER 1949) SCHUURMAN 1977
    Incertae sedis
      Camarozonosporites sp.
      Leptolepidites major COUPER 1958
      Leptolepidites psarosus NORRIS 1953

Gymnospermae
Pteridospermatophyta
  Caytoniales
    Vitreisporites pallidus (REISSINGER 1950) NILSSON 1958
Cycadophytina
  Bennettitales
    Otozamites mundae (MORRIS 1850) TEIXEIRA 1948
    Otozamites sp.
  Cycadales? Bennettitales?
    Cycadopites sp.
Coniferophytina
  Cheirolepidiaceae
    Brachyphyllum sp.
    Corollina sp.
  Araucariaceae
    Callialasporites div. sp.
  Cheirolepidiaceae? Taxodiaceae?
    Pagiophyllum sp.
  Taxodiaceae
    Cerebropollites macroverrucosus (THIERGART 1949) SCHULZ 1967
  Protopinaceae
    Prototaxodioxylon sp.
  Pinaceae
    bisaccate pollen
  Ginkgoatae
    Ginkgoaceae?
      Cycadopites follicularis WILSON & WEBSTER 1946

Charophyta
  Porocharaceae GRAMBAST 1962
    Porochara MÄDLER 1952
      Porochara westerbeckensis MÄDLER 1955
      Porochara raskyae MÄDLER 1955
      Porochara fusca var. *minor* MÄDLER 1955
  Characeae RICHARD ex. AGARDH 1824
    Mesochara ?sp. 1 SCHUDACK 1993

Animalia

Ostracoda
  Cypridacea BAIRD 1845
    Cetacella MARTIN 1958
      Cetacella armata MARTIN 1958
      Cetacella inermis MARTIN 1958
      Cetacella striata (HELMDACH 1971)
  Cytheracea BAIRD 1850
    Theriosynoecum BRANSON 1935
      Theriosynoecum wyomingense (BRANSON 1935)
    Bisulcocypris (ROTH 1933)
      Bisulcocypris pahasapensis (ROTH 1933)
    Timiriasevia MANDELSTAM 1947
      Timiriasevia guimarotensis SCHUDACK 1998
    Poisia HELMDACH 1971
      Poisia bicostata HELMDACH 1971
      Poisia clivosa HELMDACH 1971
    Dicrorygma (Orthorygma) POAG 1962
      Dicrorygma (Orthorygma) reticulata CHRISTENSEN 1965
  Darwinulacea BAIRD & NORMAN 1889
    Darwinula BRADY & ROBERTSON 1870
      Darwinula leguminella (FORBES 1855)

Bivalvia LINNÉ 1758
  Pteridomorphia BEURLEN 1944
    Pterioida NEWELL 1965
      Isognomon rugosus (MÜNSTER 1835)
  Palaeoheterodonta NEWELL 1965
    Unionoida STOLICZKA 1871
      "Unio" cf. *alcobacensis* CHOFFAT 1885

Gastropoda CUVIER 1797
  Caenogastropoda COX 1959
    Littorinimorpha GOLIKOV & STAROBOGATOV 1975
      Rissooidea GRAY 1847
        Vitrinellidae BUSH 1897
          Teinostoma H. & A. ADAMS 1854

Cerithiimorpha GOLKOV & STAROBOGATOV 1975
  Cerithioidea FERUSSAC 1819
    Procerithiidae COSSMANN 1905
      Cryptaulax TATE 1869
Heterostropha FISCHER 1885
  Archaeopulmonata MORTON 1955
    Ellobioidea H. & A. ADAMS 1855
      Ellobiidae H. & A. ADAMS 1855
        Melampoides YEN 1851

Chondrichthyes HUXLEY 1880
  Elasmobranchii BONAPARTE 1838
    Euselachii HAY 1902
      unnamed order
        Hybodontoidea ZANGERL 1981
          Hybodontidae Owen 1864
            Hybodus sp. (= *Hybodus* sp. aff. *polyprion* AGASSIZ?; SAUVAGE 1897-98)
          Acrodontidae CASIER 1959
            Asteracanthus biformatus KRIWET 1995
          Polyacrodontidae GLICKMAN 1964
            Polyacrodus sp. nov.
        Hybodontiformes indet.
    Neoselachii COMPAGNO 1977
      Galeomorphii COMPAGNO 1973
        Orectolobiformes APPLEGATE 1972 indet.
      Batomorphii CAPPETTA 1980
        Batomorphii sp. 1
        Batomorphii gen. et sp. nov. 2

Osteichthyes HUXLEY 1880
  Actinopterygii KLEIN 1885
    Actinopteri COPE 1871
      Neopterygii REGAN 1925
        Pycnodontiformes LEHMAN 1966
          Pycnodontidae AGASSIZ 1833
            Macromesodon sp.
            Coelodus/Proscinetes sp.
        Halecostomi REGAN 1923
          Semionotidae WOODWARD 1890 incertae sedis
            Lepidotes sp. 1
            Lepidotes sp. 2
          Ionoscopidae LEHMAN 1966 incertae sedis
            gen. et sp. indet.
          Macrosemiidae THIOLLIÈRE 1858 incertae sedis
            gen. et sp. indet.
        Amiiformes Huxley 1861
          Caturidae OWEN 1860
            cf. *Caturus*
          Caturidae indet.
    Teleostei MÜLLER 1846
      Teleostei indet.
      incertae sedis
        Pachycormidae indet.

Amphibia LINNÉ 1758
  Allocaudata FOX & NAYLER 1982
    Albanerpetontidae FOX & NAYLER 1982
      cf. *Celtedens* McGowan & Evans 1995
  Caudata OPPEL 1811
    family incertae sedis
      cf. *Marmorerpeton* EVANS, MILNER & MUSSETT 1988
  Anura RAFINESQUE 1815
    Discoglossidae GÜNTHER 1858
      Discoglossidae indet.

Testudines BATSCH 1788
incertae sedis
  eggshells (*Testudoolithus*)
  Casichelydia GAFFNEY 1975
    Cryptodira COPE 1870
      Pleurosternidae COPE 1868
      Pleurosternidae indet.
    Pleurodira COPE 1870
      Platychelyidae BRÄM 1965
      Platychelidae indet.

Squamata OPPEL 1811
  Scincomorpha CAMP 1923
    Paramacellodidae ESTES 1983
      Becklesius ESTES 1983
        Becklesius hoffstetteri (SEIFFERT 1973)
      Paramacellodus HOFFSTETTER 1967
        Paramacellodus sp. indet.
    Scincoidea OPPEL 1811
      family incertae sedis
      Saurillodon ESTES 1983
        Saurillodon proraformis (SEIFFERT 1973)
        Saurillodon ?henkeli (SEIFFERT 1973)
        Saurillodon cf. *obtusus* (OWEN 1850)
    Scincomorpha indet.
  Anguimorpha FÜRBRINGER 1900
    family incertae sedis
      Dorsetisaurus HOFFSTETTER 1967
        Dorsetisaurus pollicidens (SEIFFERT 1973)
  "Platynota" (sensu PREGILL et al. 1986)
    family incertae sedis
      Parviraptor EVANS 1994
      Parviraptor sp. indet.

Crocodyliformes BENTON & CLARK 1988
  Mesoeucrocodylia WHETSTONE & WHYBROW 1983
    Thalattosuchia FRAAS 1902
      Teleosauridae COPE 1871
        Machimosaurus v. MEYER 1837
          Machimosaurus hugii v. MEYER 1837
    Neosuchia BENTON & CLARK 1988
      Goniopholididae COPE 1875
        Goniopholis OWEN 1841
          Goniopholis cf. *simus* OWEN 1878
  Mesoeucrocodylia incertae sedis
    Lisboasaurus estesi SEIFFERT 1973

Pterosauria KAUP 1834
"Rhamphorhynchoidea" PLIENINGER 1901
  Rhamphorhynchidae SEELEY 1870
    Rhamphorhynchinae NOPSCA 1928
      aff. *Rhamphorhynchus* v. MEYER 1847
    Rhamphorhynchinae indet.
Pterodactyloidea PLIENINGER 1901
  family incertae sedis

Dinosauria OWEN 1842
  Ornithischia SEELEY 1887
    Ornithopoda MARSH 1881
      Hypsilophodontidae DOLLO 1882
        Phyllodon THULBORN 1973
          Phyllodon henkeli* THULBORN 1973
      Iguanodontia DOLLO 1882
        Iguanodontia indet.*
  Saurischia SEELEY 1887
    Sauropodomorpha HUENE 1932
      Sauropoda MARSH 1878
        (?)Brachiosauridae RIGGS 1904
          Brachiosauridae indet.*
    Theropoda MARSH 1881
      Ceratosauria MARSH 1884
        Ceratosauria indet.*
      Tetanurae GAUTHIER 1986
        ?Allosauroidea (MARSH 1879)
          ?Allosauroidea indet.*
        Coelurosauria HUENE 1914
          Coelurosauria incertae sedis
            cf. *Compsognathus* WAGNER 1861
              cf. *Compsognathus* sp.*
          ?Tyrannosauridae OSBORN 1905
            Stokesosaurus MADSEN 1974
              Stokesosaurus sp.
            Tyrannosauridae indet.*
          Maniraptora GAUTHIER 1986
            Deinonychosauria COLBERT & RUSSELL 1969
              Dromaeosauridae (MATTHEW & BROWN 1922)
                Dromaeosaurinae MATTHEW & BROWN 1922
                  Dromaeosaurinae indet.*
                Velociraptorinae BARSBOLD 1983
                  Velociraptorinae indet.*
              (?)Troodontidae GILMORE 1924
                (?)Troodontidae indet.*
            Aves LINNAEUS 1758
              cf. *Archaeopteryx* V. MEYER 1861
                cf. *Archaeopteryx* sp.*
      Theropoda incertae sedis
        cf. *Richardoestesia* CURRIE, RIGBY & SLOAN 1990
          cf. *Richardoestesia* sp.*
        aff. *Paronychodon* COPE 1876
          aff. *Paronychodon* sp.*

Mammaliaformes ROWE 1988
  Docodonta KRETZOI 1946
    Docodontidae SIMPSON 1929
      Haldanodon KÜHNE & KRUSAT 1972
        Haldanodon exspectatus KÜHNE & KRUSAT 1972

Mammalia LINNAEUS 1758
  Multituberculata
    Paulchoffatiidae HAHN 1969
      Paulchoffatia KÜHNE 1961
        Paulchoffatia delgadoi KÜHNE 1961
        Paulchoffatia sp. A
      Meketibolodon HAHN 1978
        Meketibolodon robustus (HAHN 1978)
      Guimarotodon HAHN 1969
        Guimarotodon leiriensis HAHN 1969
      Plesiochoffatia HAHN & HAHN 1999
        Plesiochoffatia thoas (HAHN & HAHN 1998)
        Plesiochoffatia staphylos (HAHN & HAHN 1998)
        Plesiochoffatia peparethos (HAHN & HAHN 1998)
      Xenachoffatia HAHN & HAHN 1998
        Xenachoffatia oinopion HAHN & HAHN 1998
      Kuehneodon HAHN 1969
        Kuehneodon dietrichi HAHN 1969
        Kuehneodon guimarotensis HAHN 1969
        Kuehneodon simpsoni HAHN 1969
        Kuehneodon dryas HAHN 1977
        Kuehneodon uniradiculatus HAHN 1978
      Kielanodon HAHN 1987
        Kielanodon hopsoni HAHN 1987
      Meketichoffatia HAHN 1993
        Meketichoffatia krausei HAHN 1993
      Pseudobolodon HAHN 1977
        Pseudobolodon oreas HAHN 1977
        Pseudobolodon krebsi HAHN & HAHN 1994
      Henkelodon HAHN 1977
        Henkelodon naias HAHN 1977
      Bathmochoffatia HAHN & HAHN 1998
        Bathmochoffatia hapax HAHN & HAHN 1998
    Albionbaataridae KIELAN-JAWOROWSKA & ENSOM 1994
      Proalbionbaatar HAHN & HAHN 1998
        Proalbionbaatar plagiocyrtus HAHN & HAHN 1998
  Dryolestida PROTHERO 1981
  Dryolestidae MARSH 1879
    Dryolestes MARSH 1878
      Dryolestes leiriensis MARTIN 1999
    Krebsotherium MARTIN 1999
      Krebsotherium lusitanicum MARTIN 1999
    Guimarotodus MARTIN 1999
      Guimarotodus inflatus MARTIN 1999
  Paurodontidae MARSH 1887 (including Henkelotheriidae KREBS 1991)
    Henkelotherium KREBS 1991
      Henkelotherium guimarotae KREBS 1991
    Drescheratherium KREBS 1998
      Drescheratherium acutum KREBS 1998
  Amphitheriida PROTHERO 1981
    Zatheria McKENNA 1975
      family incertae sedis
        gen. et sp. nov.

---

* Dinosaur taxa only known from teeth.